METAL-AIR BATTERIES

This monograph outlines the principles of operation of metal-air batteries, and the current published state-of-the-art of metal-air battery technology. Information has been collected from a variety of publications, mainly released during the period 1967–70.

The text includes a general description of the advantages and disadvantages of metal-air batteries when compared to conventional batteries, and outlines the historical development of zinc-air systems. The principles underlying the air electrode are explained, and the basis for the selection of the metal electrode is outlined.

Several different types of metal-air batteries are described. Some are already in production, some are in development, and some are still in the research phase. Primary, secondary, and mechanically rechargeable zinc-air systems are described, together with magnesium, aluminium, cadmium and sodium-air units.

A brief bibliography includes suggestions for further reading.

Other titles in this series

CE/1 MICROBIOLOGICAL CORROSION
G H Booth

CE/2 SOLVENT TREATMENT OF COAL
W S Wise

CE/3 THE FLUIDISED COMBUSTION OF COAL
D G Skinner

CE/4 CARBONISATION OF COAL
J Gibson and D H Gregory

CE/5 THE PRINCIPLES OF GAS EXTRACTION
P F M Paul and W S Wise

CE/7 FUEL CELLS
D P Gregory

Other Chemical Engineering titles available

TL/CE/1 BORON
A G Massey and J Kane

All published by Mills & Boon Limited

M & B Monograph CE/6

General Editor: J Gordon Cook, PhD, FRIC

Metal-Air Batteries

D P Gregory, BSc, PhD

Assistant Director of the Engineering Research Division,
Institute of Gas Technology, Chicago

Mills & Boon Limited

London

First published in Great Britain 1972
by Mills & Boon Limited, 17–19 Foley Street,
London W1A 1DR

ISBN 0 263 51712 8

Set in 11/12 pt. Monotype Baskerville,
printed by photolithography, and bound
in Great Britain at The Pitman Press, Bath

CONTENTS

1. INTRODUCTION 7

2. HISTORY 10

3. PRINCIPLES OF OPERATION 17
(1) The Air Electrode, or Cathode 17
(2) The Metal Electrode, or Anode 21
(3) The Electrolyte 26

4. TYPES OF METAL-AIR CELL 28
(1) Primary Cells 28
(2) Secondary Cells 28
(3) Mechanically Rechargeable Cells 28

5. RECENT DEVELOPMENTS IN METAL-AIR BATTERIES 32
(1) Mechanically Rechargeable Cells 32
(2) Primary Cells 49
(3) Secondary Cells 56

6. APPLICATIONS OF METAL-AIR BATTERIES 66
Conclusions 68

Appendix 71

Bibliography 77

Index 79

ACKNOWLEDGEMENTS

The author wishes to thank the Management of Energy Conversion Ltd., for permission to publish the material in this monograph, and to thank Mr. T. M. Fry for many helpful suggestions in editing the text.

1. Introduction

An electrochemical cell contains essentially two electrodes and an electrolyte. One electrode is capable of being oxidised electrochemically, or is in contact with an oxidisable material; in the course of the oxidising reaction electrons are liberated, rendering this electrode negatively charged. The other electrode undergoes a reduction reaction which consumes electrons, leaving the electrode positively charged. A conventional battery positive electrode contains a metallic oxide component, which is reduced to a lower oxidation state when the battery operates. When all the oxide has been converted, the electrode is said to be discharged and the cell is either discarded or recharged electrically.

Metal oxides, however, are not the only reducible materials which have been used in positive electrodes. In principle, almost any oxidising agent will serve. Materials which have no electronic conductivity may be used, so long as they are oxidising agents and can be brought into contact with a suitable conducting electrode. Oxygen itself can be employed as the oxidising agent; it is reduced, in effect, to water in the electrode reaction. Since oxygen is relatively unreactive, a catalyst must be employed to assist the electrode reaction. An oxygen electrode thus consists of a conducting component, at which oxygen and the electrolyte are brought into contact with a suitable catalyst. Being a gas, oxygen can be fed continuously to the electrode itself, and the product water removed. Thus, an oxygen electrode can in principle run forever, so long as there is a supply of oxygen. An oxygen electrode's capacity is limited by the size of the oxygen storage tank, and not by the amount of material which can be stored in the structure of the electrode.

There is an inexhaustible supply of oxygen in the earth's atmosphere. An electrode designed to utilise the oxygen of air as its reactive material has an infinite capacity. The weight and volume of the porous catalytic electrode structure required to function as the air electrode are relatively low compared to those of a high capacity metal-oxide electrode. Air electrodes can be made to run continuously at higher current densities (current per unit area) than most metal oxide electrodes.

The air electrode, therefore, has three distinct advantages over the conventional battery positive electrode:

1. Infinite charge capacity.
2. Low weight, independent of capacity.
3. Capability of sustained high current operation.

These benefits are gained only at the expense of some disadvantages. Air electrodes tend to be more sophisticated structures than "consumable" metal oxide electrodes, and are therefore more expensive in low capacity units. The electrode must be open to air and electrolyte during operation; as the electrode is porous, it exposes the electrolyte to the air. The concept of a "sealed cell" cannot, therefore, be applied to an air battery, and evaporation of the electrolyte can occur, limiting the battery life. The electrochemical potential of the oxygen-hydroxyl couple is lower than that of most metal oxidation couples, and the operating voltage of an air cell is thus lower than that of most conventional cells.

The air electrode, therefore, has three potential disadvantages over conventional battery positive electrodes:

1. More expensive (especially in low capacity cells).
2. Limited life in use.
3. Low voltage.

There are many fields where the listed advantages

outweigh the disadvantages, and because the metal-air cell, or "air-depolarised battery" is inherently so attractive, a great deal of research and development has been carried out on these systems.

2. History

Leclanché publicly described his now famous cell in 1868, by which time 20,000 of them had already been installed in railway signalling and telegraph lines. This cell consisted of an amalgamated zinc negative electrode, and ammonium chloride electrolyte, and a positive electrode made from a mixture of manganese dioxide (the consumable material) and carbon powder (for conductivity). Leclanché noted that if the electrolyte only half-filled the jar, and the upper portion of the manganese dioxide/carbon mixture was only moist, rather than immersed in the electrolyte, the performance of the cell was improved. Clearly the cell was benefiting from the action of atmospheric air at the porous carbon electrode; Leclanché had, unwittingly, constructed the first metal-air cell! In 1879 Maiche followed up this observation by adding platinum to the carbon powder, and omitting the manganese dioxide. He retained the zinc and ammonium chloride of the Leclanché cell, and thus demonstrated the first true zinc-air battery.

In 1932, an air-depolarised zinc battery with caustic alkali electrolyte was described by Heise and Schumacher. This battery used a porous carbon block electrode, which was prevented from becoming completely flooded with the electrolyte by impregnation with paraffin wax. This type of cell was subsequently produced by the Union Carbide Company in the U.S.A. as a two-cell battery; each cell contained a pair of solid zinc electrodes, amalgamated with mercury to prevent self-discharge, one electrode being on each side of a wet-proofed porous carbon block whose upper edge was open to the air above the surface of the sodium hydroxide electrolyte (see Fig. 1). Access of air to the internal pore structure of the carbon block was thus somewhat restricted, and the cells were capable of operating only at relatively low rates. The carbon pores were impregnated with paraffin wax to prevent their becoming flooded, and this was effective only as long as the current density was kept low. It was considered that if the oxygen in the pores and in contact with the electrolyte became exhausted, the electrolyte would flow in to occupy the space, thus flooding the electrode.

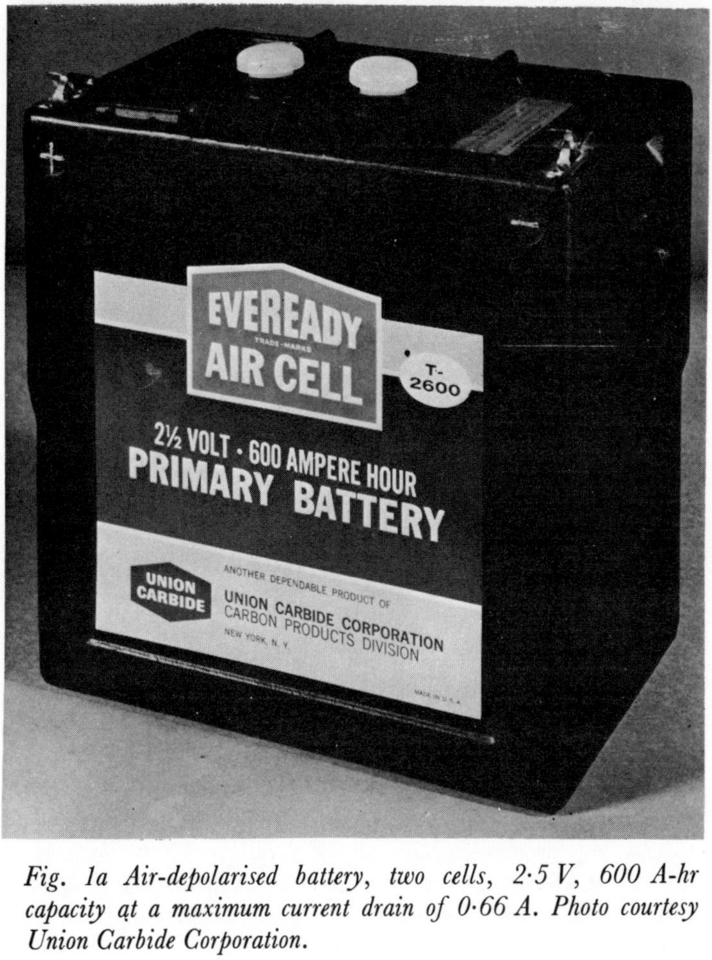

Fig. 1a Air-depolarised battery, two cells, 2·5 V, 600 A-hr capacity at a maximum current drain of 0·66 A. Photo courtesy Union Carbide Corporation.

Fig. 1b Air-depolarised batteries of 1000 A-hr and 3000 A-hr capacities.

Cells capable of delivering 0·65 A for 1,000 hr, and 2 A for about 300 hr were used for radio batteries and for railway signalling batteries respectively. Similar cells were in production and use in European countries during the same period.

In about 1950, the National Carbon Company, U.S.A., developed a small air-depolarised hearing-aid battery. This unit, shown in Fig. 2, contained a perforated zinc

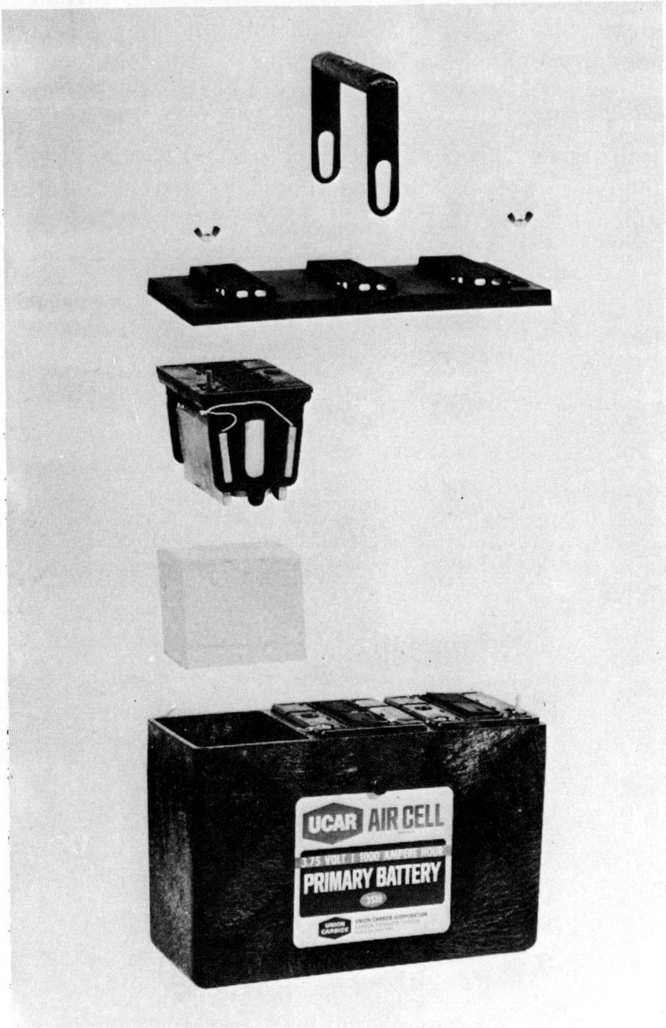

Fig. 1c Internal components of the air depolarised battery.

electrode; there were two thin wet-proofed carbon air electrodes, one on each side of the zinc. The electrolyte was immobilised, as in all "dry" cells. The cell was assembled in a vinyl plastic case, and the air access holes were covered by adhesive tape until the battery was put into service. The battery, which had a capacity of about 0·4 A-hr, was the first commercial "dry-cell" primary zinc-air battery

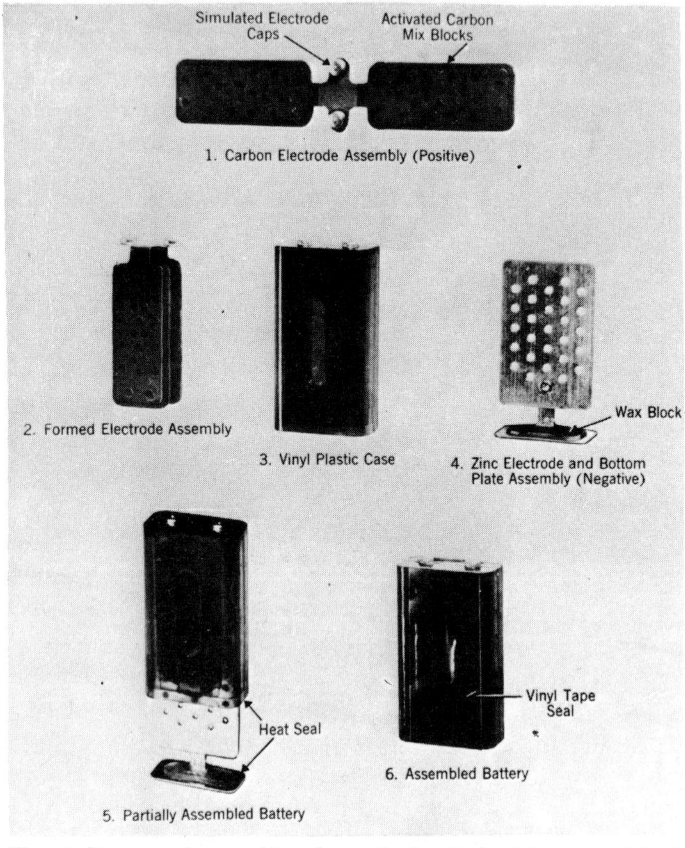

Fig. 2 Parts and assembly of an air-depolarised hearing aid A battery, Eveready No. 1005-E. Photo courtesy Union Carbide Corporation.

The cell used for railway signalling, shown in Fig. 3, could be obtained in a "renewable" form. A new electrode assembly and calcium hydroxide regenerative material were provided. The calcium hydroxide precipitated the zinc in solution, regenerating the electrolyte, and the used zinc anodes were replaced with new ones. This was probably the first instance of a "mechanically rechargeable" metal-air cell.

Fig. 3 Eveready air-depolarised signal cell, type AC 500, capacity 500 A-hr. Photo courtesy Union Carbide Corporation.

Towards the end of the 1950s, the development of fuel cells gathered impetus, and a great deal of research and development was carried out after about 1960 on air electrode structures for room-temperature alkaline electrolyte cells. As "fall-out" from this work, high current density air electrodes became available, and they were incorporated in metal-air cells, especially zinc-air cells, from about 1965 onwards.

The problems involved in development of the fuel cell are formidable. Not least of these are the choice of fuel and availability of hydrogen, and the management of water removal systems. Both of these problems are eliminated by the use of zinc as the "fuel", giving zinc-oxide as the product. Thus the complexity of the system is reduced, and the reliability is increased.

There was an upsurge of interest in metal-air (particularly zinc-air) research and development, during the latter part of the 1960s. This has not yet resulted in a new family of high-rate zinc air cells being produced on a commercial scale, but these cells have found uses in military applications, and wider development is imminent.

3. Principles of Operation

(1) THE AIR ELECTRODE, OR CATHODE

Oxygen can be reduced electrochemically to form hydroxyl ions, as follows:

$$4e + O_2 + 2H_2O \rightleftharpoons 4OH^-$$

The reaction takes place readily, provided (1) that there is an adequate catalyst for the reaction, and (2) that there is an electronic conductor available to yield the electrons required. Thus, the reaction can occur only at the surface of a conducting, catalytic electrode which is in contact with both oxygen and electrolyte. The thermodynamic reversible potential of this process is 1·23 V positive to a hydrogen electrode in the same electrolyte, so it would be expected that an electrode in contact with oxygen and electrolyte would stabilise at this potential. In practice, potentials of only about 1·00 V or 1·05 V are observed, and when the current is drawn, further loss of voltage (or polarisation) occurs, so that a plot of electrode potential versus current would appear as in Fig. 4. Most other

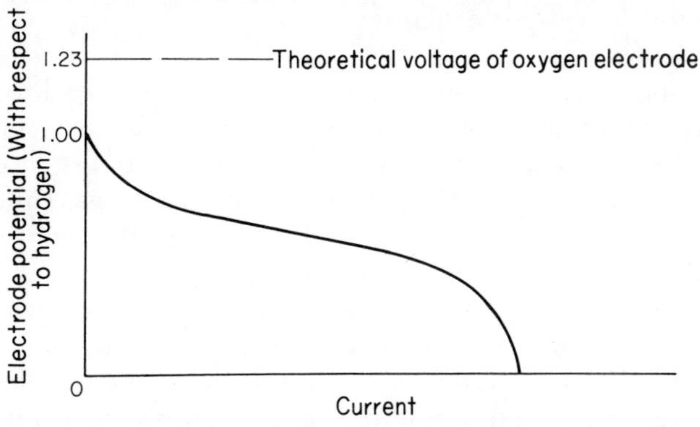

Fig. 4 Typical current-voltage curve for an oxygen electrode.

electrode systems also lose voltage when current is drawn, but they usually exhibit the theoretical potential for that particular system when no current flows.

The loss of voltage of the air electrode at open circuit (zero current) is due to the fact that the electrode reaction does not progress by the direct route represented by:

$$4e + O_2 + 2H_2O \rightarrow 4OH^- \qquad (1)$$

but by a route involving the intermediate production of a peroxide-type of ion, as represented by:

$$2e + O_2 + H_2O \rightarrow O_2H^- + OH^- \qquad (2)$$

The O_2H^- ion is decomposed chemically at the electrode surface

$$O_2H^- \rightarrow OH^- + \tfrac{1}{2}O_2 \qquad (3)$$

and the oxygen can re-enter the original reaction (2).

The reversible thermodynamic potential for reaction (2) is 0·83 V, whereas for reaction (1) it is 1·23 V. It is suggested that a mixed reaction occurs involving both reactions (1) and (2) above, so that the potential of the electrode stabilises somewhere between the two, but there is still controversy among electrochemists as to the exact nature of the reaction, and the true reason for the low observed electrode potential.

Having already forfeited some efficiency by this loss of potential at zero current, the cell suffers a further loss as current is drawn. Initially, this loss arises from concentration differences which are set up both in the gas supply to the electrolyte interface and in the electrolyte itself, and also as a result of the slowness of the electron transfer reaction. Finally, the current reaches a value where complete starvation of oxygen occurs, and a "limiting current" is observed. This corresponds to the condition where oxygen is diffusing to the reaction site at its maximum rate, and this current therefore cannot be exceeded.

The entire science of air electrode design is centred around the reduction of these polarisation voltage losses to minimum values, and the maximising of the limiting current value.

Oxygen, from the air, is available in the gas phase. The electrolyte is usually in the liquid phase, whilst the electron conductor, needed for completion of the reaction, is normally a solid. The reaction, therefore, has to occur in the vicinity of three adjacent phases and the design of electrode must be such as to stabilise the interfaces between the three phases and to maximise the overall area of contact.

Oxygen or air electrodes are usually in the form of a porous, conducting solid, into which can penetrate both the electrolyte and the gas. There are several methods of stabilising the gas-liquid interface within the pores, and they depend upon capillary action. By wet-proofing some of the pores, the contact angle is raised, and electrolyte cannot penetrate. This wet-proofing is effected by paraffin wax, or by P.T.F.E. or similar hydrocarbon or fluorocarbon materials. Alternatively, if two distinctly different pore sizes are employed, it is possible to arrange the conditions so that the liquid floods the fine pores, but is kept out of the coarse pores by a positive gas pressure. The former principle is normally employed for metal air batteries.

The electrode must be provided with an active catalyst to promote the electron transfer reaction. This catalyst must be electronically conducting and resistant to corrosion by the electrolyte. Fuel cell electrodes conventionally employ platinum group metals as catalysts, but these are precluded from being used in metal-air cells as they can migrate to the metal electrode, where they cause self-discharge. Other non-noble metals must be used, and most successful metal-air cells have used carbon or silver, in various forms, as catalysts.

If a metal-air cell is to be used without any auxiliary moving parts, a compressed air supply can be discounted, and means must be provided in the electrode to prevent the electrolyte from leaking out through the electrode pores. This, too, is accomplished by a wet-proofing technique, and some air battery electrodes are clad on the air side with a thick layer of porous water-repellent plastic, such as P.T.F.E. or polyethylene.

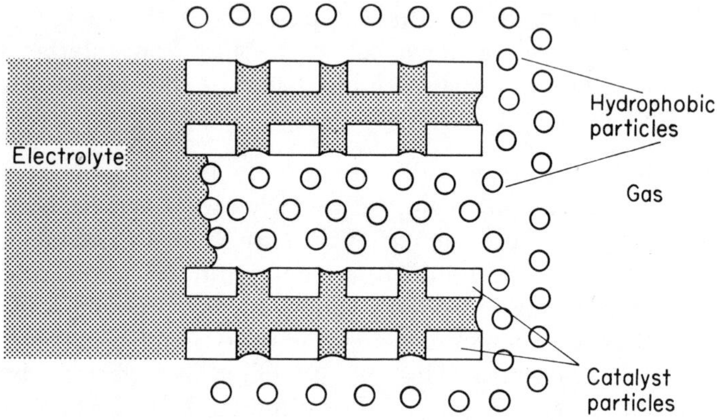

Fig. 5 Schematic structure of a hydrophobic electrode.

An example of an air electrode is based on a porous nickel sheet, which acts as a structural member and current collector. One face of the sheet is coated with a mixture of catalyst particles and P.T.F.E. particles. An interlocking pore structure is thus obtained; electrolyte enters the nickel and penetrates the pores within the catalyst particles. Air can make its way into the pores between the P.T.F.E. particles, and it is thus in contact with the entire surface of the catalyst particles. An alternative to this structure uses a woven nickel screen, rather than a porous sheet, with a layer of porous P.T.F.E. on the air side to prevent liquid leakage. Other variations include the use of porous carbon sheets (which are cheaper and bulkier) instead of nickel sheets, and porous carbon sheets having some of their pores rendered wet-proof with wax, P.T.F.E., or polyethylene. Figure 5 shows a schematic cross-section of a hydrophobic interlocking pore structure air electrode.

(2) THE METAL ELECTRODE, OR ANODE

The anode of a metal-air cell is usually similar to a conventional anode of a primary or secondary battery system. Many metals may be oxidised electrochemically to form either metal oxides, or metallic ions in solution. For example, in the case of zinc in acidic solutions, zinc ions are formed

$$Zn \rightarrow Zn^{++} + 2e$$

whilst in alkaline solutions, zinc oxide is produced

$$Zn + 2OH^- \rightarrow ZnO + H_2O + 2e$$

In the presence of excess alkali, the product is soluble zincate ion

$$Zn + 4OH^- \rightarrow ZnO_2^{--} + 2H_2O + 2e$$

Since the original metal is an electronic conductor, and since most electrochemical oxidation reactions of metals are fast, it is not necessary to supply conductors or catalysts; the metal itself can act as its own electrode. During the reaction, however, the electrode is either dissolved or changed chemically, and the quantity of electricity that can pass (the "capacity" of the electrode) is determined by the amount of metal present.

Most metal electrodes exhibit their theoretical electrochemical potentials when no current is drawn, and polarise when a current flows. This polarisation is due mainly to the electronic resistance of the electrode, and to the slowness of diffusion of reactants and products in the electrolyte. Since the electrode geometry changes radically during the discharge of the battery, the factors contributing to polarisation change with time, so that the polarisation tends to increase during discharge. This is in contrast to the air electrode, which is invariant, and whose polarisation is therefore constant with time.

The design of metal electrodes is concerned with the optimisation of the structure to give minimum polarisation, especially when almost completely oxidised, and to prevent corrosion of the metal from occurring when no current is passing.

Several practical metal-air systems have been considered, and in all of them the aim is to oxidise a metal electrochemically to its oxide by the use of atmospheric oxygen. In the choice of metal, a compromise is sought between the ease of oxidation, the stability of the metal in contact with electrolyte, the weight required for a given energy release, cost and availability, and the completeness with which oxidation can be accomplished. Zinc is the usual choice; it is recognised as having the most useful overall characteristics, but the choice is clearly influenced by the intended application of the battery.

The "infinite capacity" of the air electrode offers a potentially lightweight cell, and the choice of metal is therefore influenced strongly by the available energy per unit weight, or energy density. Note must be taken of the voltage of the battery as well as the A-hr capacity, for it is the "W-hr" figure which is relevant to energy density.

Table 1 lists the theoretical energy content of various metals assuming their oxidation in an electrochemical cell. The weight of the oxygen is ignored, as is the weight of electrolyte, container, current connections, etc.

The rate at which this energy can be released is important, but no logical or predictable method is available for classifying the metals into an order of priority. Generally speaking, high rates of reaction can be obtained from those metals which can be formed into high surface area components, and which do not form a protective oxide layer when oxidised electrochemically.

Table 1 Theoretical Energy and Potential Characteristics of Various Metal/Oxygen Couples

CELL	THEORETICAL POTENTIAL V	ENERGY DENSITY W-hr/lb
Lithium-Oxygen	3·44	6020
Aluminium-Oxygen	2·70	3645
Magnesium-Oxygen	3·09	3090
Zinc-Oxygen	1·65	614
Iron-Oxygen	1·28	556
Cadmium-Oxygen	1·21	261
Lead-Oxygen	1·59	186
For Comparison		
Silver-Zinc Battery	1·86	254
Lead-Acid Battery	2·14	112

If the primary cell is required to have long shelf life, the metal electrode must be stable when in contact with the electrolyte. Most metal-air systems under consideration use aqueous electrolytes and it can be assumed that those metals with an electrochemical potential below that of hydrogen will not react with water and evolve hydrogen, and will therefore be stable in storage. Unfortunately, this criterion also establishes the voltage of the cell, and metals with electrode potentials below hydrogen will give low voltages (below 1·23 V) when paired with an air electrode. Some metals which are above the hydrogen potential can be inhibited from corrosion in aqueous systems by using chemical inhibitors, and a compromise is therefore sought between storage characteristics, cell voltage, and cost of inhibitors. Table 2 shows electrode potentials of some common metals, indicating which are available for consideration as primary anodes.

A primary cell uses the metal anode as its energy source,

Table 2 Electrochemical Potentials of some Electrode Reactions in Alkaline Solutions

	REACTION	VOLTAGE
Unstable in aqueous solution	$Ca + 2OH^- \rightarrow Ca(OH)_2 + 2e$	3·03
	$Mg + 2OH^- \rightarrow Mg(OH)_2 + 2e$	2·69
	$Al + 4OH^- \rightarrow H_2AlO_3^- + H_2O + 3e$	2·35
Stable in aqueous solution with inhibitors	$Zn + 2OH^- \rightarrow Zn(OH)_2 + 2e$	1·25
	$Fe + 2OH^- \rightarrow Fe(OH)_2 + 2e$	0·88
*	$H_2 + 2OH^- \rightarrow 2H_2O + 2e$	0·83
Stable without inhibitors	$Cd + 2OH^- \rightarrow Cd(OH)_2 + 2e$	0·81
	$Ni + 2OH^- \rightarrow Ni(OH)_2 + 2e$	0·72
	$2Cu + 2OH^- \rightarrow Cu_2O + H_2O + 2e$	0·358
	$2Ag + 2OH^- \rightarrow Ag_2O + H_2O + 2e$	—0·344
	$4OH^- \rightarrow O_2 + 2H_2O + 4e$	—0·401

* Voltages are "Standard Electrochemical Potentials", the voltage of the electrode in contact with 1 N concentrations of reactants and products, and compared to a normal hydrogen electrode in acidic electrolyte. In alkaline solution, the hydrogen potential is 0·83 V, and the tendency of a metal to displace hydrogen from solution is measured by the voltage difference between the above voltages and 0·83 V.

and the cost of energy is related directly to the cost of metal. For high power batteries, this becomes a significant criterion. Zinc is an expensive "fuel" on which to run a T.V. set or an electric drill, but it is quite acceptable for low power devices such as transistor radios and electric wrist watches. Table 3 shows the raw material costs (1970) of various metals, converted to power costs in electrical units.

Table 3 Raw material costs of various metals expressed in electrical units, based upon 100% efficient conversion

METAL	COST £ PER TON	COST NEW PENCE PER KWH
Aluminium	257	2
Magnesium	368	5
Zinc	121	11
Cadmium	3,700	630
Iron	35	2·7

In order to obtain a high yield of energy from a metal electrode, steps must be taken to prevent the surface becoming blocked off by the metal oxide product as the discharge takes place. For example, a solid zinc plate will operate only as long as there is an excess of alkaline electrolyte to dissolve the product. As the zincate in solution reaches saturation, a blocking layer of oxide forms on the surface of the zinc, and the cell ceases to function. The weight and volume of the electrolyte required in a high capacity battery render the whole concept rather unattractive. However, if the zinc is provided in a high surface area, highly porous form, almost complete conversion to solid zinc oxide may be achieved, the electrode life being limited by the decrease in the conductivity of the solid compact.

In general terms, steps must be taken to retain the

conductivity of the metal electrode as it nears complete conversion to oxide. This can be done by the incorporation of an inert metallic screen current collector, or by the addition of an inert conducting component, e.g. graphite powder, to the formulation of the electrode material itself. Again, care is taken to avoid using too great an amount of inert material which would have the effect of lowering the energy density obtained.

A secondary cell must be capable of electrical recharge, and the metal must therefore be capable of being electrochemically reduced in the presence of electrolyte. This requirement rules out some high-electrical-potential metals, such as magnesium or aluminium, for use in secondary batteries, with aqueous electrolyte. In general, the available metals for secondary cells are the same as those identified in Table 2 as being capable of storing their energy in contact with aqueous electrolytes.

(3) THE ELECTROLYTE

The function of the electrolyte is to act as a conductor, enabling the ionic species produced at one electrode to travel to the other, where they are consumed. Usually, water is also involved in the electrode reactions and this, too, is carried across the cell by the electrolyte. Because the electrolyte is not an electronic conductor, it prevents the passage of electronic current between the two electrodes.

The air electrode puts the electrolyte in direct contact with the air, and the choice of electrolyte is thus restricted. Volatile materials will tend to evaporate, and must either be replenished or supplied in such quantity as not to limit the life of the battery. In general, the use of aqueous electrolytes can be tolerated, since the provision of excess water is economically feasible. Volatile organic and non-aqueous electrolytes, however, can seldom be considered.

The air electrode produces hydroxyl ions, and its potential depends upon the concentration of hydroxyl ions in solution. If OH^- ions are allowed to build up locally, the potential changes rapidly, causing severe polarisation. Buffering the solution helps to reduce the pH change, or hydroxyl ion concentration change, when OH^- ions are introduced; it thus reduces the polarisation and allows higher currents to be drawn. Acids and alkalies are good buffer solutions; for this reason, the most satisfactory air electrodes operate in highly alkaline or highly acidic electrolytes. Each has its disadvantages; alkaline electrolytes become contaminated with carbon dioxide from the air, whilst acid electrolytes are more corrosive to inexpensive catalysts and current collectors used in air electrodes. In practice, it is more usual to accept the disadvantages of the alkaline electrolyte. Some metal-air systems use nearly-neutral aqueous electrolytes, such as sodium chloride or potassium carbonate, but they are limited to low current density operation.

4. Types of Metal-Air Cell

Metal-air cells fall into three principal classes: (1) primary cells, (2) secondary cells, and (3) mechanically rechargeable cells.

(1) PRIMARY CELLS

A cell which, once discharged, is of no further value and is discarded, is called a primary cell. Most of the early zinc-air cells were primary cells; they were capable of operating for very long periods at low current drains. A successful primary cell must be relatively cheap to buy, and have a long storage life; it should offer a convenient "energy package" of either low weight or low volume, or both.

(2) SECONDARY CELLS

A cell which, once discharged, can be reactivated by passing a current through the cell in the reverse direction is called a secondary cell. Unlike a conventional secondary lead-acid or nickel-cadmium cell, however, a secondary metal-air cell has one electrode of infinite capacity, which can be neither completely discharged nor overcharged. On charging, the air electrode evolves oxygen into the atmosphere.

(3) MECHANICALLY RECHARGEABLE CELLS

A third type of metal-air cell is known as the "mechanically rechargeable cell" or the "replateable cell". Once the cell has been discharged, the spent metal electrode (now oxidised) is removed and discarded, and a new one is introduced. This change may also include a fresh charge of electrolyte, but the essential feature is that the air electrode, which can never "run down" is used many times over.

The spent anode removed from a mechanically rechargeable cell could, in principle, be sent to a central processing station where it would be reduced chemically or electrochemically, returning it to its original state. Though this is inconvenient in practice, it offers another alternative type of use.

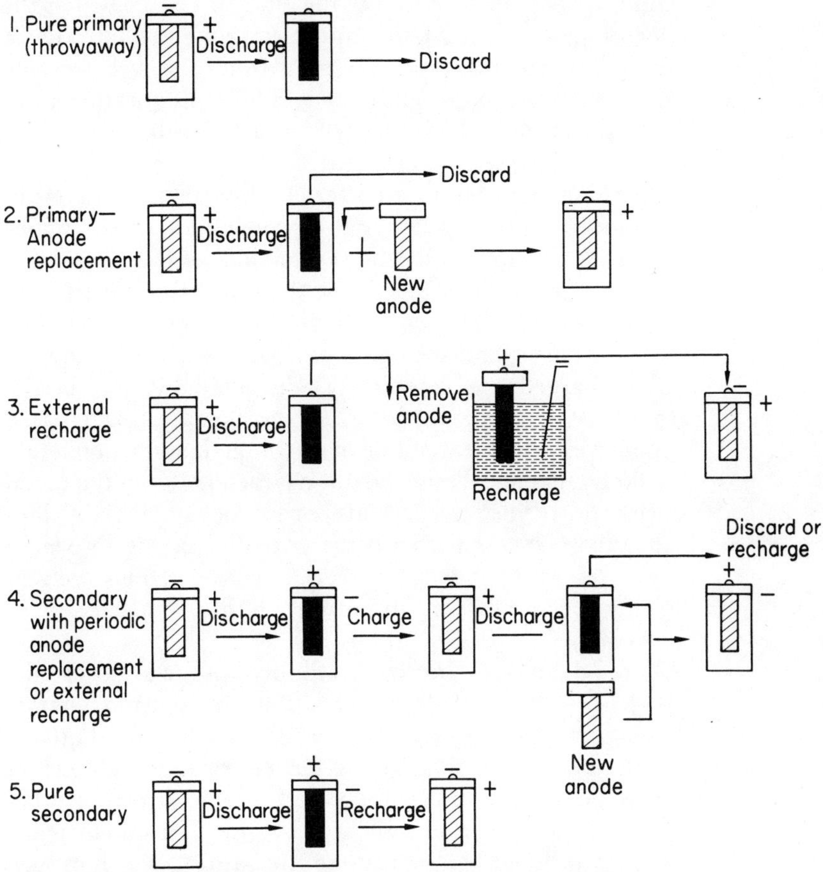

Fig. 6 Alternative operating schemes for zinc-air batteries.

Similarly, a "hybrid" of the mechanically rechargeable cell and the true secondary can be considered, in which electrical recharge is limited to a few cycles. Periodically, a mechanical replacement of the metal electrode is carried out, extending the recharge life of the battery. The various modes of operation are summarised schematically in Fig. 6.

Mechanical recharge can be carried out intermittently or continuously. In a version of the zinc-air cell known as the "dry tape cell", fresh zinc and electrolyte are introduced in tape form, the tape emerging from the cell carrying zinc oxide and spent electrolyte. This configuration is, in fact, a continuously fed battery or a fuel cell.

The secondary metal-air battery is of particular interest in a number of respects. Most conventional secondary batteries are limited by the reversibility of the "positive", or metal oxide electrode. Thus, the lead oxide electrode of the lead-acid battery normally limits the rate of discharge, rate of charge, and the depth of discharge. If the electrode is discharged too completely, the available number of cycles is reduced. In contrast, the metal-air cell has a "positive" which cannot be over-charged, and can operate at high currents. Thus, the full characteristics of the metal electrode ("negative" plate) can be exploited. This advantage is coupled with the potential saving in weight by eliminating the heavy oxide electrode. This weight-saving is more evident in batteries of high capacity.

On the debit side, the "irreversibility" of the air electrode, evidenced by its failing to exhibit the thermodynamic theoretical voltage, results in a loss of efficiency both on discharge and on charge. A secondary metal-air cell is likely to exhibit more than half a volt difference between its charge and discharge voltage. This inefficiency leads to a rise in temperature of the battery, which in turn increases the evaporation of the water of the electrolyte.

This is important both on discharge, where air flows past the electrodes, and on charge, when a stream of water-saturated oxygen is evolved. A secondary metal-air battery will therefore require an air pump, cooling system, and provision for water make-up. These extra components represent a disadvantage, which may, however, be more than counterbalanced by the advantages of lighter weight and higher rates of charge and discharge.

5. Recent Developments in Metal-Air Batteries

(1) MECHANICALLY RECHARGEABLE CELLS

ZINC-AIR CELLS

Recent work has concentrated largely on the development of mechanically rechargeable zinc-air and zinc-oxygen systems which are intended primarily for military use in communications devices. In these applications, relatively high rates and relatively large amounts of energy are required, and light-weight systems are desirable.

A typical cell construction takes the form of a flat sheath, with a pair of air electrodes facing each other, and a central flat zinc anode, as shown in Fig. 7. The cell is constructed in such a way that the zinc anode, with its cap and seal, may be withdrawn when the cell has been discharged, and replaced with a new one. In order to develop the required voltage, several cells are stacked in series. In operation, this stack of cells is clamped together so that the anodes are held rigidly in place; they are prevented from touching the air electrodes by porous insulating separators placed inside each cell. The clamping device is released after discharge, so that the anodes may be withdrawn easily.

Figure 8, which shows a single cell containing two cathodes, is called a "Bi-Cell". This particular design was developed by Leesona-Moos Laboratories[1] in the U.S.A., but it is typical of many similar versions. The anode is a porous zinc plate, moulded with a plastic "lid" or cap, and jacketed in a porous paper separator (see Fig. 9). A conducting current collector embedded within the anode makes contact with a terminal in the cap of the anode.

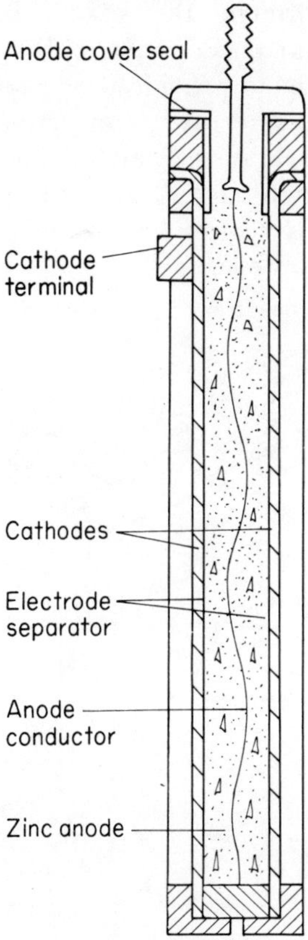

Fig. 7 Sheath-type cell construction.

When the anode is inserted into the sheath of the cell, this terminal enters a socket in the cell case which is connected electrically to the cathodes of the next cell along the battery, thus linking the cells together in series.

A series of cells is stacked to form a battery, each cell being separated by a so-called "intercell separator". This fits between two adjacent cathodes, and provides an important control function in the battery's operation. Air is allowed to diffuse through the spaces between the cathodes, and for low currents or short pulses, the normal diffusive flow is sufficient to provide the required oxygen feed. At sustained moderate rates, convective air flow is initiated by the waste heat of the battery, and this flow is needed to provide enough oxygen. At exceptionally high rates, forced air flow is required to remove waste heat and prevent the cell from overheating.

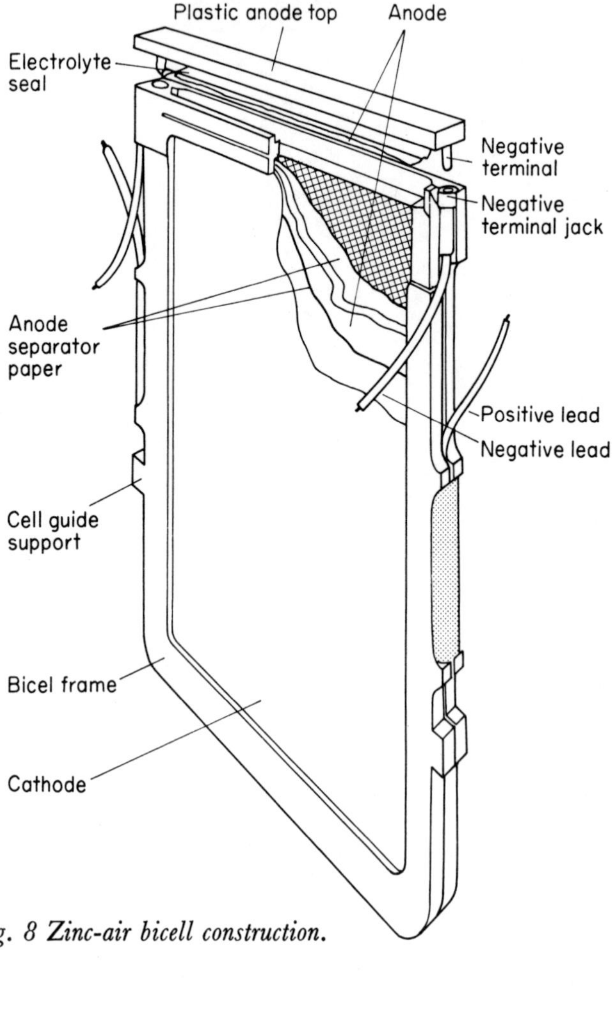

Fig. 8 Zinc-air bicell construction.

The critical points of design of the intercell separator are (a) its thickness, which controls the amount of oxygen contained in the cavity to provide pulses, (b) its porosity, which controls the air flow rate and optimises between oxygen supply, temperature control, and water evaporation rates, and (c) its composition, plastic being employed at low rates, and metal separators in batteries designed for high rates to facilitate heat removal by conduction. The separators also provide a support to keep the flexible air electrodes pressed against the anode; in some designs, they allow for the growth or swelling of the anode during discharge, maintaining an even pressure, and thus an even contact between the anode, electrolyte, and cathode.

The most serious failure mechanism reported by Leesona-Moos is related to water loss. Since the cathode is permeable to air, it can also permit the transpiration of water vapour. The water loss is enhanced at high ambient temperatures, at high current drain, and at low ambient relative humidities. In certain designs this can be minimised by incorporating excess electrolyte, which may be located, for example, at the sides of the cell. Water lost through the face of the cathode is replenished by gravity and capillary feed through the anode structure. The presence of excess electrolyte is also of value in reducing the battery operating temperatures. Evaporative cooling by the water transpired through the cathode surface is a significant factor.

The capacity of the zinc-air system is related directly to the proportion of utilizable zinc in the anode structure. Most zinc-air designs attempt to maximise this amount. In anodes of 0·16 to 0·25 inches thickness, zinc utilization figures from 50 to 90 per cent have been reported, depending on the thickness, plate density, and current drain. The design of the anode is therefore a compromise between efficiency, strength, and overall size (area/thickness ratio).

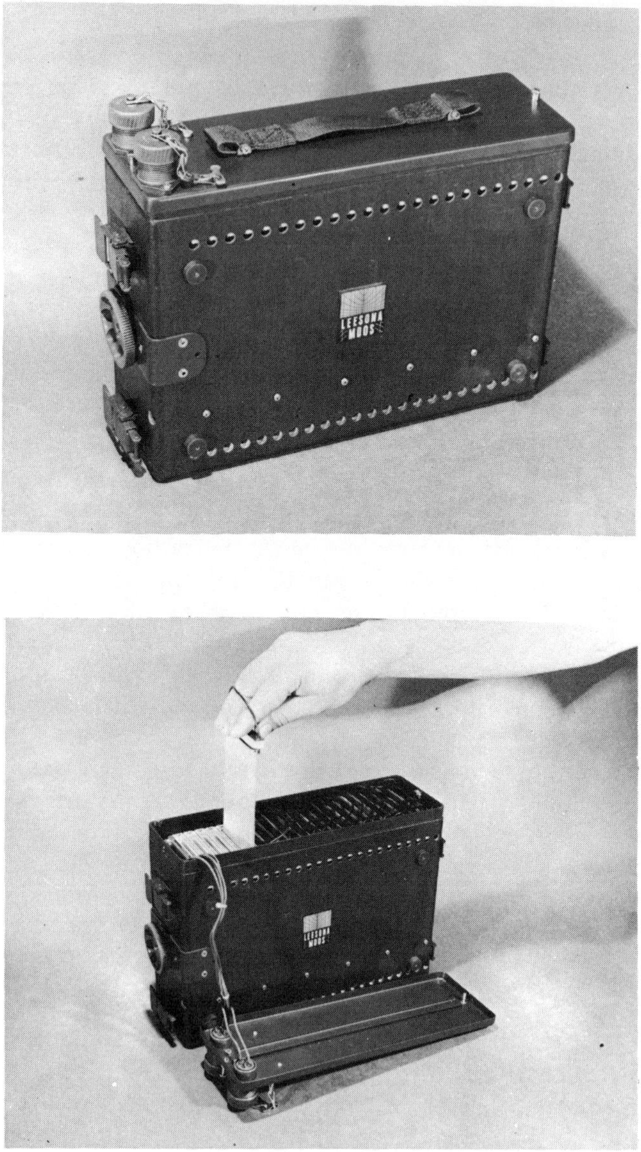

Fig. 9 Mechanically rechargeable, 24 V, 25 A-hr, high rate zinc-air battery. Photo courtesy Leesona Moos Laboratories.

Two mechanically rechargeable zinc-air systems, developed in the U.S.A. for military applications, have been fairly fully described[6]. One is a 24 V 25 A-hr system, and the other is a 24 V 48 A-hr system. The former is described in Tables 4 and 5, which show characteristics of the high-rate and low-rate systems. The high-rate system incorporates an air blower to provide forced air cooling; it is shown in Fig. 9.

These batteries are recharged by removing the spent anodes and discarding the electrolyte. Water is poured into the empty cell cases until it reaches the top. Each anode in the recharge kit is then removed from its sealed aluminium foil pouch, and inserted slowly into the water-filled cell. During manufacture, the anodes have been impregnated

Table 4 Characteristics of a 24 V Low Rate Mechanically Rechargeable Zinc-Air Battery

ELECTRICAL		
Current drain	2 A	4 A
A-hr capacity @ 70°F	30 A-hr	22 A-hr
Nominal voltage	28·0 V	27·0 V
Open-circuit voltage	31 V	31 V
Specific volume	2·3 W-hr/in^3	1·6 W-hr/in^3
Energy density (to 22 V cut-off)	78 W-hr/lb	55 W-hr/lb

PHYSICAL	
Length: 11½in	Volume: 368in^3
Height: 8in	Activated weight: 10¾lb
Width: 4in	

RECHARGE	
Recharge kit dimensions	5¾in H × 6in W × 9½in L
Weight	5lb
Recharge time	10min
Battery life	>50 recharges

with electrolyte (KOH) and dried, prior to wrapping. After the battery has been "refuelled", the cover is replaced, and full power is immediately available. Recharging can be accomplished in less than 10 minutes.

A similar battery has been developed in the U.K. by Energy Conversion Ltd.* This is a nominal 12 V 24 A-hr system, the characteristics of which are shown in Table 6. The battery illustrated in Fig. 10 contains eleven cells connected in series. It will produce currents of up to 3 A; a modified battery incorporating a fan to blow air through it has delivered a continuous current of 20 A.

The battery, without a blower, warms up to about 50°C on full load, and the discharge curve is very flat, as shown in Fig. 11. Most metal-air batteries display this characteristic of maintaining an almost constant voltage until they are almost completely discharged. The voltage then falls rapidly.

Because most of the zinc in the battery is oxidised usefully, and the other components of the battery are relatively light, an energy density of nearly 100 W-hr/lb is achieved.

The battery is capable of operating, though with reduced capacity, to temperatures as low as —40°C. Figure 12 shows that this zinc-air battery has a significantly higher energy/weight ratio than all conventional primary battery systems at these sub-zero temperatures.

After discharge, the clamps are slackened off, and the anodes can be withdrawn from the cells. Residual electrolyte is emptied out. In this case, refuelling is carried out by measuring into each cell a fresh charge of potassium hydroxide electrolyte. A new set of anodes is inserted; the battery is clamped up, and is ready for use

* D. P. Gregory. Design Engineering, Feb. 1970, p. 67.

again. The recharge operation is silent and takes only a few minutes; as it does not use pure water, it can be carried out under freezing conditions.

Two methods of electrolyte replacement during recharging have been described which avoid handling the alkaline electrolyte. In one method[6] the porous zinc anodes are impregnated at the factory with the appropriate electrolyte, then dried and packed in waterproof packs. The battery is activated by filling it with

Table 5 Characteristics of a 24 V, 25 A-hr, High Rate Mechanically Rechargeable Zinc-Air Battery

ELECTRICAL*	
A-hr capacity @ 70°F	25 A-hr
Nominal voltage†	25·3 V
Open-circuit voltage	25·5 V
Maximum current drain‡	15 A
Specific volume	1·3 W-hr/in^3
Energy density (to a 22 V cut-off)	50 W-hr/lb
PHYSICAL	
Length: 11½in	Volume: 460in^3
Height: 10in	Activated weight: 12·5lb
Width: 4in	
RECHARGE	
Recharge kit dimensions	5¾in H × 6in W × 9½in L
Weight	5lb
Recharge time	10min
Battery life	>50 recharges

* Rated for cyling load 90 per cent low rate (30 W) 10 per cent high rate (300 W).

†Voltage regulator integral in design.

‡ Higher currents (> 15 A) can be sustained for pulsing applications.

clean water and inserting the anodes. Problems include the stability of zinc in contact with solid, highly concentrated alkali, penetration of water and CO_2 through the storage bag, and the effects on the battery of the heat-release on dissolution of the electrolyte.

In the second method, a reservoir of electrolyte solution is included in a cavity within each anode. Inserting the anode in the battery breaks a seal, allowing the electrolyte to penetrate the anode pores.

Several different systems are still under development, and it seems probable that the method of recharge will be chosen to fit the particular application for which the battery is designed.

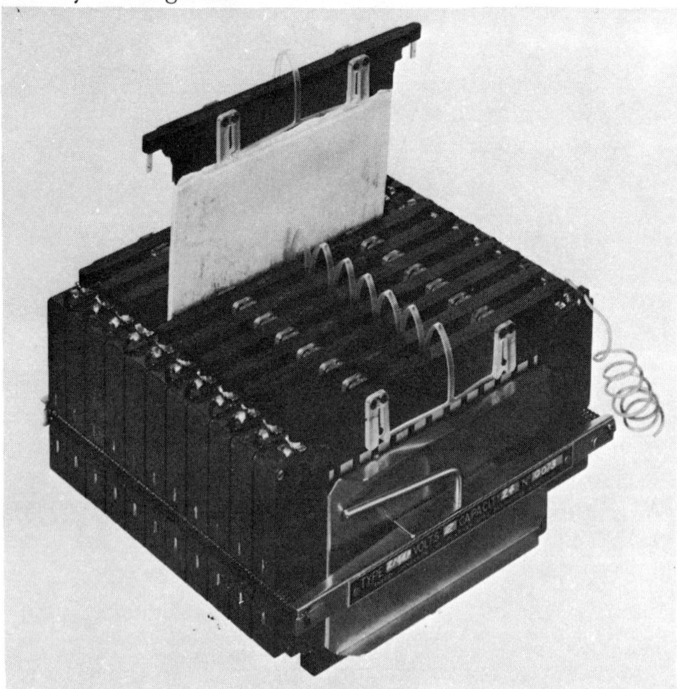

Fig. 10 ECL 12 V 22 A-hr battery. Photo courtesy Energy Conversion Ltd.

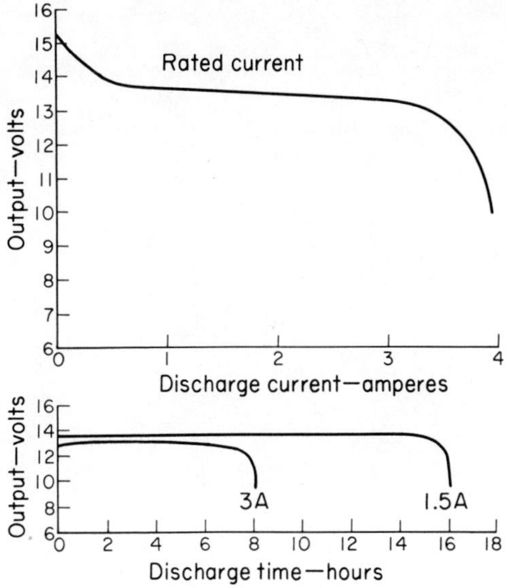

Fig. 11 Discharge curves for ECL 24 A-hr zinc-air battery.

Table 6 Characteristics of Energy Conversion Ltd., Zinc-Air Battery

Specification for type ZRO1/A/1.5

Physical data

Weight	$3\frac{1}{2}$lb
Height	4in
Width	$6\frac{3}{8}$in
Length	$6\frac{3}{8}$in

Performance at 20°C

V	13·5
A	1·5
W	20
A-hr	24 (to 10 V)
W-hr	315

Electrolyte Potassium-hydroxide solution

Re-energised by replacement of its discharged anodes.

Operating temperature range −40°C to +55°C

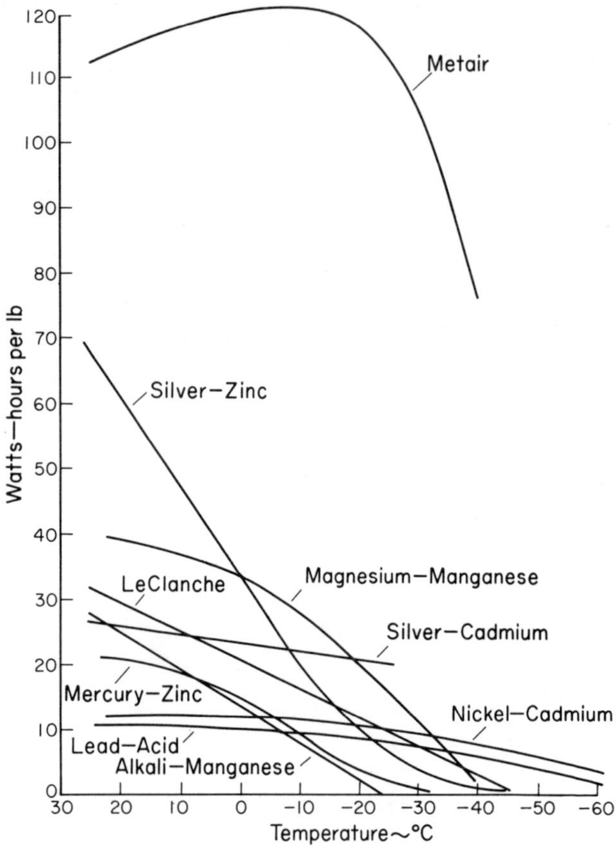

Fig. 12 Manner in which energy-weight ratios of various battery systems vary with temperature.

ALUMINIUM-AIR CELLS

An aluminium-air cell should exhibit a voltage of 2·71 V and theoretically should produce 3600 W-hr per pound of metal consumed (though simultaneous consumption of water will half this figure).

This high theoretical potential and energy density of aluminium make it attractive as an anode for a metal-air cell, and its relatively low cost adds to this attraction. However, in practice, in aqueous electrolytes, aluminium exhibits a voltage far below its theoretical value; this may be due to the formation of a thin skin of aluminium oxide on the surface. The slow diffusion of aluminium through this skin may be speeded up by the addition of a trace of tin to the metal. Aluminium tin alloys (0·2 per cent tin) exhibit higher voltages, but are still not 100 per cent efficient.

Aluminium anodes do not require a highly alkaline electrolyte and will discharge satisfactorily in neutral electrolytes such as potassium chloride, ammonium chloride, and aluminium chloride, but air electrodes require a highly buffered electrolyte in order to achieve high current.

An aluminium-air cell using a neutral electrolyte exhibits poor voltage efficiency and a low limiting current. A strongly alkaline electrolyte acts as a buffer, and eliminates the latter problem. However, it is found that even lower open circuit voltages are produced, due to the spontaneous evolution of hydrogen by corrosion of the aluminium by the alkali. Attempts at inhibiting this corrosion or hydrogen evolution, for instance, by introduction of chromate ions, have resulted in deterioration of the air electrode as well. Hydrogen evolution takes place on open circuit, and the cell cannot be stored in an activated state.

The poor voltage efficiency of the cell results in the release of very large amounts of heat during discharge, since only half the chemical energy is converted to electrical energy. The need for cooling mechanisms is an unattractive feature of aluminium-air cells.

These problems have restricted the development of useful aluminium-air cells. However, the possibility of construct-

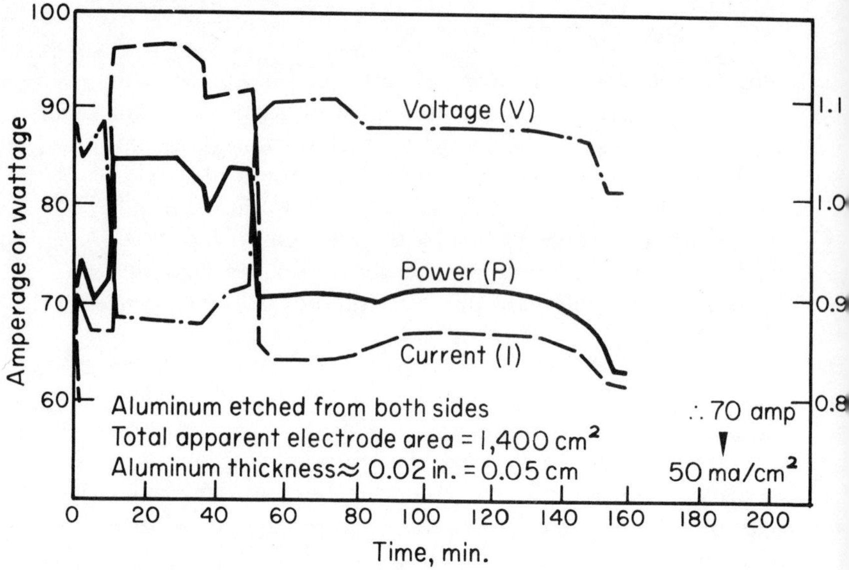

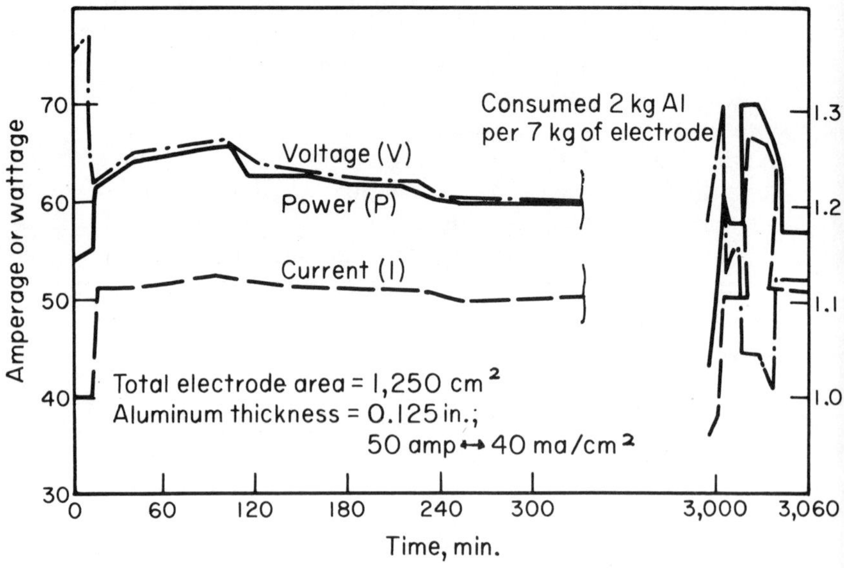

Fig. 13 Top: Typical performance of a 70 W aluminium-air battery with thin anodes. Kinks in curves correspond to load adjustments (June, 1966).
Bottom: Performance of the same battery with thick anodes of smaller area (March, 1967).

ing a cell which gives a very high energy density, a high voltage per cell, and a cheap source of packaged electricity, is an attractive goal to the research worker.

Zaromb Research Corporation[9] reported in 1967 a primary aluminium-air battery capable of producing 70 W, operating with a circulating KOH electrolyte and aluminium alloy anodes. The electrolyte is removed when the battery is out of use, to prevent corrosion, and it is intended that the product $Al(OH)_3$ should be removed from the electrolyte stream and replacement aluminium anodes inserted. Performance data are given in Fig. 13.

MAGNESIUM-AIR CELLS

Magnesium-air cells have an even higher theoretical voltage (3·09 V) than aluminium-air cells. Although this is advantageous from a theoretical standpoint, this voltage is far above that of hydrogen and it is difficult to imagine magnesium being stable in an aqueous electrolyte without self-discharge, or hydrogen evolution, taking place. However it is possible to reduce the self-discharge rate to acceptable limits by the use of a magnesium-aluminium-zinc alloy. This feature has stimulated a great deal of research on magnesium-air cells, especially by the U.S. General Electric Co.[8]

The electrolyte usually employed is a sodium chloride solution of about 7 per cent concentration. Magnesium tends to self-discharge to form hydrogen and magnesium hydroxide, but the latter forms a protective film and prevents further attack. This film, although beneficial with respect to the storage properties of the cell, creates a problem when the cell is to be discharged, for it must be removed before a large current can be drawn. As the drawn current is increased, the film is spontaneously disrupted by local concentration gradient effects, but the magnesium anode will not respond to sudden demands for pulses of current. A compromise must be sought between the resistance of the film to self-discharge, and the rapidity with which it can be broken down when required. The properties of the film can be varied by alloying the magnesium with various other metals.

For general purposes, the "AZ" series of alloys are used, the cheapest and most common being AZ 31B. (3·0 per cent Al, 1·0 per cent Zn, 0·2 per cent Mn, 0·005 per cent Ni, 0·005 per cent Fe, 0·040 per cent Ca.)

The open circuit voltage of a magnesium-air cell should be 3·09 V. In practice, only about 1·6 V is obtained, and the voltage falls to about 1·3 V on drawing a useful current. Most of this loss comes from the magnesium electrode; the entire loss, representing a poor efficiency, results in a large release of heat when the battery is operating. Although the electrodes are capable of delivering over 100 mA/cm^2, cooling is required above about 30 mA/cm^2 if the battery is not to "run away" thermally.

This high heat release also promotes evaporation of the electrolyte; this may be overcome only by providing an adequate reservoir, which negates part of the energy-density advantage. Additionally, although solid sheet metal electrodes are used, the hydroxide falls to the bottom of the cell as a sludge, and room must be left for this in the battery design, adding to the weight and volume of the unit.

Figure 14 shows a diagram of a "Mag-air" single cell and Fig. 15 shows components of a battery. The similarity to the zinc-air battery is apparent. The characteristics of a 24 V battery described by the U.S. General Electric Co.[8] are shown in Table 7.

Figure 16 shows a typical magnesium-air performance curve.

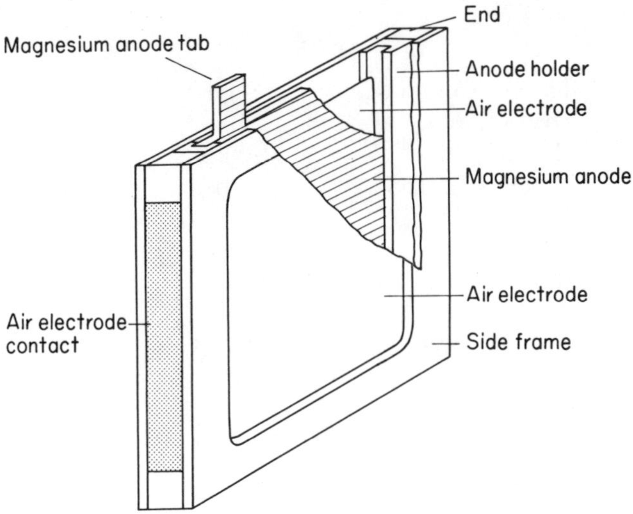

Fig. 14 Basic magnesium-air single cell design.

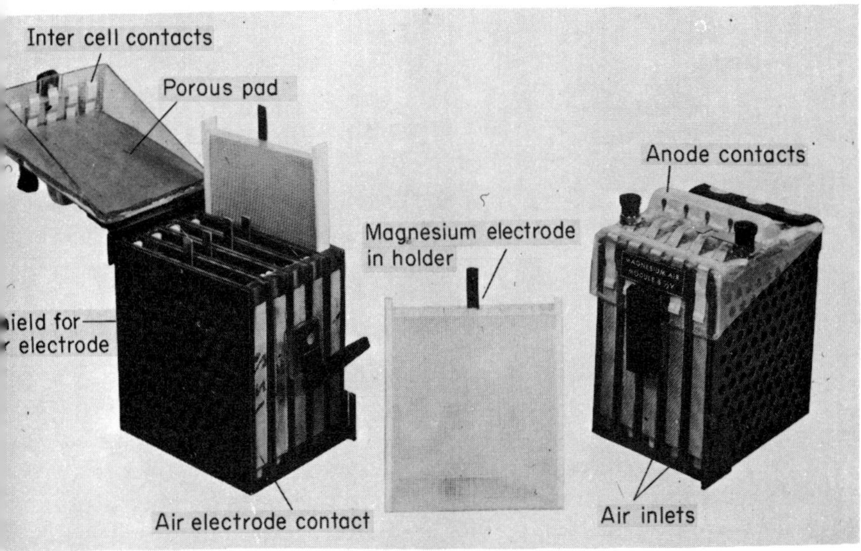

Fig. 15 Five-cell magnesium-air battery components. Photo courtesy General Electric Company.

Table 7

Battery Component	Design Value
Dry weight	4·25lb
Wet weight	11·00lb
Refuel Kit (Dry)	1·75lb
Dimensions	4·25 × 8·0 × 10·0in
Operating V	24
Refuel Cycle	12hr
A-hr/charge	26·4
W-hr/lb (wet)	58·8

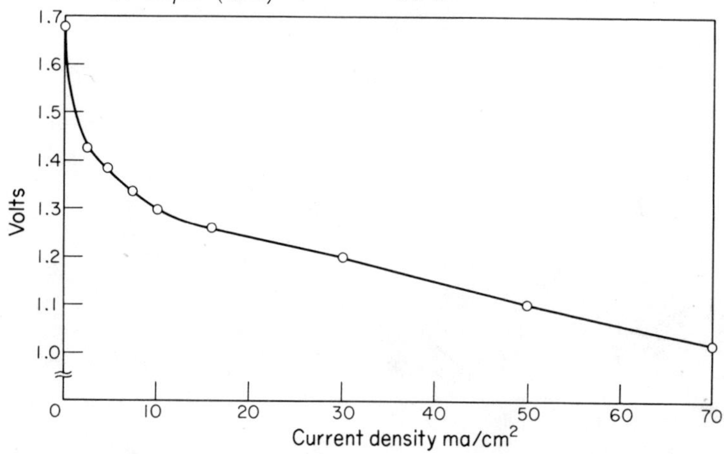

Fig. 16 Typical performance curve of a magnesium-air cell.

Advantages of the magnesium-air battery over the zinc-air battery include simpler, cheaper metal anodes, and the use of salt water as electrolyte, which is less hazardous than alkali. However, the magnesium-air battery fails to achieve the promised advantage of higher energy density (less than $\frac{1}{3}$ the value of zinc-air is obtained), and it probably has a shorter activated storage life. Furthermore, it cannot deliver pulses of current. However, the advantages of the magnesium-air battery, especially in energy cost (replacement anodes at £1·20 ($3·00) per KWH are predicted), could make it commercially significant in the future.

(2) PRIMARY CELLS

ZINC-AIR CELLS

One of the more recent successful primary cell developments is the "manganese-alkali" cell.* This is a Leclanché type cell which uses potassium hydroxide electrolyte instead of ammonium chloride. It allows a more advantageous use of the zinc anode, which is provided in the form of a porous, highly active compact. The compatibility of this form of anode with potassium hydroxide solutions has been ably demonstrated in these commercial cells.

It is a relatively small conceptual step to replace the manganese dioxide electrode with an air electrode. This offers a considerable reduction in weight, for the air electrode is light and independent of capacity. The weight saving applies especially to higher capacity units, making a primary zinc-air battery attractive in principle.

The main problem confronting development of primary zinc-air batteries has been the cost of the air electrodes because, in fuel cell technology, platinum usually has been used as the catalyst. As cheaper catalysts are developed, and cheaper electrode concepts become available, this obstacle is being removed.

Another problem is associated with the fact that the air electrode must be exposed to the air and to the electrolyte, and absorption of carbon dioxide will occur through the electrode pores. This can be minimised, for example, by using a shutter to reduce the access of air to the battery, and by covering the air access holes (e.g. with adhesive tape) to eliminate contact during storage. A shutter might be incorporated in the device in which the battery is used; it would be operated by the on-off switch, isolating the battery from the air when current is not actually being drawn.

* Kirk–Othmer, Encyclopaedia of Chemical Technology, 2nd. Ed., **3,** p. 117.

In order to prevent self-discharge of the zinc during storage, a poisoning agent is added to the zinc to inhibit the evolution of hydrogen. Zinc will normally be attacked according to reactions of the following type:

$$Zn + 2OH^- \rightarrow ZnO + H_2O + 2e \qquad (4)$$

$$2H_2O + 2e \rightarrow H_2 + 2OH^- \qquad (5)$$

overall $$Zn + H_2O \rightarrow ZnO + H_2$$

The electrode will take up the equilibrium potential of reaction (4). As this is well below the hydrogen evolution potential, reaction (5) will proceed spontaneously. If, however, the metal surface is coated with a "poison" for reaction (5), or the metal itself is rendered non-catalytic, reaction (5) is inhibited or slowed down, so that neither reaction can occur (reaction (4) can proceed only if there is a means of removing the electrons produced).

Mercury is a well known inhibitor of hydrogen evolution, and it alloys freely with zinc. Amounts of mercury varying from 2 to 10 per cent are commonly used for this purpose in many zinc-containing primary cells, and zinc-air cells are no exception. For economic reasons, however, the amount of mercury must be kept to a minimum; 1 per cent of mercury costs about the same as the total amount of zinc.

Early zinc-air cells used solid plates of zinc, coated with mercury, as anodes. They were placed in cells which contained a plentiful supply of potassium hydroxide electrolyte. When current is drawn, zinc is oxidised to form zinc oxide, which in turn is dissolved in the electrolyte to form zincate ions.

$$Zn + 2OH^- \rightarrow ZnO + H_2O + 2e$$

$$ZnO + 2OH^- \rightarrow ZnO_2^{--} + H_2O$$

The reaction proceeds so long as there is some zinc remaining, and as long as the external circuit is allowed to remove the electrons formed, and the supply of OH^- ions is maintained by the air electrodes.

If, however, the solution becomes saturated with zincate ions, which are only sparingly soluble, zinc oxide is deposited on the zinc plate as an impervious, passivating layer, and the reaction ceases. The cell can be "reactivated" by addition of extra electrolyte, or by replacing the electrolyte; in general, however, this type of cell is designed to contain sufficient electrolyte to allow all the zinc to react. This results in a rather heavy and bulky battery, and does not take advantage of the potential weight saving allowed by the air electrode.

In order to overcome this penalty, modern zinc-air batteries have been developed which use highly porous zinc anodes, made by compacting small zinc particles. A high area-to-volume ratio is achieved, and this type of electrode can be converted almost entirely to solid zinc oxide without passivation, since the oxide layer does not form an impermeable barrier to further oxidation. Zinc oxide is an insulator, and as the electrode is oxidised in use, its conductivity is reduced. In order to compensate for this, an inert current collector may be embedded within the porous zinc anode. This can be a silver, copper or brass collector, in the form of a perforated screen, rods, or coils of wire.

A penalty paid for the advantages of a high surface, highly porous anode is the increased tendency to self-discharge. Just as the anode is more "active" in use, so it is more likely to be corroded or oxidised by the electrolyte in storage, and more difficult to inhibit with respect to hydrogen evolution. Mercury addition, however, serves to prevent rapid self-discharge of the cells.

The purity of the zinc anode becomes critical in highly active anodes in storage. Small impurities can have the effect of catalysing the hydrogen evolution reaction, or giving rise to "local cell" corrosion.

Anodes, fabricated from powdered zinc of irregular particle shape to provide high surface area and porosity, are usually in the form of cylinders or flat plates. The electrolyte is stored in the cell mainly within the pores of the anode, where sufficient can be held to provide for the chemical reactions that take place. In use, however, it is found usually that the cell's life is limited by the rate at which water evaporates from the electrolyte, and in order to compensate for this an electrolyte reservoir is incorporated in the cell. This may be within the anode itself, which may be, for instance, in the form of a hollow cylinder, or the reservoir may be built into the cell body in the form of extra free volume. The size of this reservoir is designed to match the expected service conditions of the cell, and it is likely that specialised zinc-air batteries will be developed with reservoirs tailored to fit their stated application, in order to minimise weight and volume.

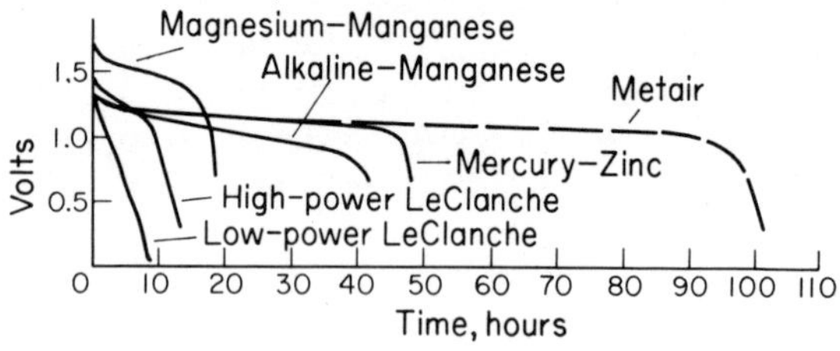

Fig. 17 Performance curves of various 'D' size batteries (250 mA drain).

Eliminating the weight and volume of the manganese dioxide from the alkaline Leclanché cell by substituting an air electrode allows a considerable saving in volume within the cell itself. This freed volume may be occupied by extra zinc, so that it is possible to build considerably more capacity into a zinc-air cell of a given size. For example, standard sized zinc-air cells are claimed to have as much as ten times the capacity of Leclanché cells of a similar size. Figure 17 shows a comparison of the capacity obtained when a number of conventional "D" size primary cells are discharged at the same current. The "D" size zinc-air cell—an experimental type not yet in production—shows significant superiority; it is illustrated in Fig. 18. This superiority is reflected in the chart of Fig. 19, which shows the capacity available from various primary cell systems of similar size.

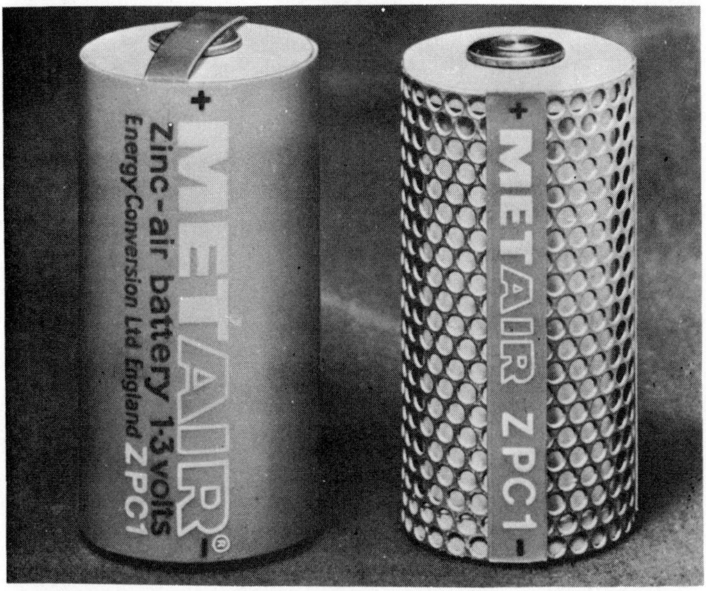

Fig. 18 Typical 'Metair' battery before and after removal of outer wrapper. Photo courtesy Energy Conversion Ltd.

With the exception of the National Carbon Co., "Ever Ready" hearing aid battery referred to earlier, no primary zinc-air dry cell had become commercially available by 1970, though market entry by several companies was considered imminent.

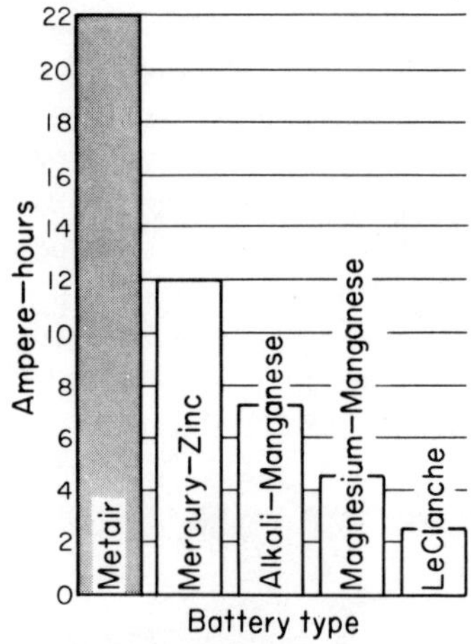

Fig. 19 Capacity of various 'D' size batteries.

*Table 8 Characteristics of Le Carbone "AR" cell Type A.D.S.O. (512)**

Dimensions, (including terminals)	$4\frac{1}{2} \times 4\frac{1}{2} \times 8\frac{1}{8}$in
Weight	7lbs
Capacity to O.C.V. on continuous discharge through 5 ohms	300 W.H
Internal Resistance	less than 0·1ohm
E.M.F.	1·4 V
Mean potential difference on discharge	above 1·0 V

Suggested Uses
Heavy Duty Line Circuits.
Signal and Point Motor Operation.
Track Circuits.
Telephone and Telegraph Circuits.
Small Automatic Telephone Exchanges.

* Le Carbone (Great Britain) Ltd., Portslade, Sussex. A. D. Battery Catalogue.

More rugged, less portable, primary zinc-air batteries known as "AD" (Air Depolarised) cells are readily available. One such cell is shown in Fig. 20 and its specification is given in Table 8. There are several similar cells available in the range 60–2500 WH, but all are limited to low current density operation. They use solid zinc anodes and an excess volume of electrolyte.

Fig. 20 "AD" high capacity dry cell, type ADSO (512). Photo courtesy Le Carbone Ltd.

(3) SECONDARY CELLS

ZINC-AIR CELLS

The comparison between the primary manganese-alkali and primary zinc-air battery holds good also in the case of the secondary battery. Secondary silver-zinc batteries are well established as lightweight secondary batteries, but suffer from severe limitations in the number of charge-discharge cycles they will withstand.

These batteries consist of a porous zinc electrode, containing mercury to prevent self-discharge, and incorporating a silver current collecting screen; this is separated by a cellulosic sheet from a silver oxide electrode, the whole being immersed in aqueous potassium hydroxide electrolyte. It is a small conceptual step to replace the silver-oxide electrode with an air electrode and achieve a secondary zinc-air battery. Such a battery would eliminate many of the cycle-life problems which are associated with the silver electrode in silver-zinc batteries, and it would exhibit an extremely useful energy density.

The situation, however, is fraught with difficulties. First, one of the basic problems of the silver-zinc battery is the tendency of zinc on recharge to plate out in "tree-like" crystals or dendrites, which penetrate the separator and cause short-circuits. Separators capable of preventing such penetration are under development, but the most successful types are not stable in warm or hot alkali. The air electrode, being less efficient than the silver electrode, will generate more waste heat; the battery will run hot, thus degrading cellulose-type separators.

Even if this problem is overcome, another problem remains. After many recharge cycles, the zinc electrode tends to change its shape, becoming thicker at the bottom and around the edges. This effect, known as "slumping" or "shape-change" in silver zinc batteries, can be minimised by suitable design of the anode, but it cannot be eliminated. It is also minimised by restricting the depth of discharge allowed for the zinc. If, however, only say 25 per cent of the zinc is allowed to oxidise, the available energy density of the battery is greatly diminished and most of the advantages of light weight are lost.

A great deal of research is being carried out in an attempt to overcome these problems, in both silver-zinc and zinc-air batteries, but no commercial secondary zinc-air cell has yet (1971) become available.

The stability of the air electrode during charge presents another problem to be overcome in any secondary metal-air battery. Under charging conditions, the air electrode potential rises about 0·5 V (or more) above the reversible oxygen potential, and oxygen is evolved within its structure. This causes severe corrosion and erosion problems. Conventional air electrode materials, such as silver and carbon, are rapidly oxidised under these conditions, and begin to go into solution. Moreover, the liberation of oxygen bubbles at a high rate within the porous structure can disrupt the carefully tailored pores and spoil the subsequent behaviour of the electrode on discharge.

The so-called "three electrode cell" has been considered by General Motors Corporation,* and others as a way of overcoming this problem. In addition to the reversible zinc electrode and the air electrode, this cell has a third "charging" electrode in the form of a woven screen or perforated metal which is placed between the other two. On discharge, the outer two electrodes are used, and the third electrode is idle. On charge, the connections are arranged so that current is passed between the zinc electrode and the charging electrode, which simply

* J. J. Postula *et al.*, "On the Use of Third Electrodes in a Secondary Zinc-Air Battery". *Energy Conversion,* **10,** No. 2, 1970, pp. 45–49.

evolves oxygen. The air electrode is idle, and remains undamaged.

Apart from adding bulk and weight to the cell, the system suffers from the great disadvantage that two sets of "bus bars" are needed in a multi cell battery, or individual switching must be carried out on each cell. This adds considerably to cost, weight and unreliability.

An interesting type of cell, designed to overcome the secondary zinc-air battery problem has been developed by the General Atomics Division of Douglas-Astropower[(7)] in the U.S., and by Joseph Lucas Ltd. in the U.K. The cell consists of a series of steel plates, on which is deposited a coating of zinc, and a series of porous nickel air electrodes, through which a flow of air is allowed to bubble.

During discharge, air from a pressurised air plenum seeps through the porous nickel cathodes and bubbles into a KOH electrolyte. The zinc oxide formed at the anodes begins to precipitate as a finely divided solid. The electrolyte is continuously circulated through the cell, and sweeps out this zinc oxide into a separator, where it is collected as a relatively dry sludge. The electrolyte flow serves also to sweep out the air and nitrogen from the cell, and it acts as a coolant. After leaving the cell, the stream passes through the air separator, a heat exchanger, and the zinc oxide separator, and is then returned to the cell.

During charge, a reverse reaction takes place. A potential is applied across the cell while the electrolyte circulates. Zinc from the zinc oxide is replated on the backing sheet, and pure oxygen is evolved from the surface of the porous nickel electrodes. The now "depleted" electrolyte continuously dissolves zinc oxide from the storage point outside the cell.

The cells and stack, containing no stored reactants other

than metallic zinc, can be quite compact and relatively light, but the system requires several pieces of equipment in addition to the cell stack, including the air and zinc oxide separators, heat exchanger, pump and electrolyte reservoir.

During discharge, the voltage varies with power demand in a manner similar to that of most battery systems. The voltage variation during discharge is relatively flat until nearly full discharge is reached, because the zinc oxide produced at the zinc electrodes is removed continuously. A diagram of the system is shown in Fig. 21 and the cell voltage characteristics are shown in Fig. 22.

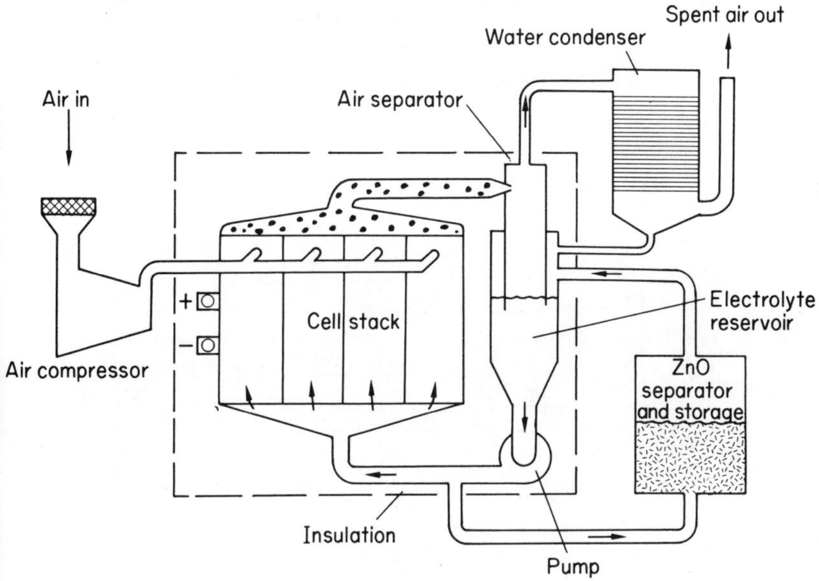

Fig. 21 Schematic of a zinc-air battery system.

The cell is stated to have an extremely fast charging rate, and in units of up to 14 kW-hr which were produced experimentally in 1966, energy densities of 50 W-hr/lb were achieved. For larger units, up to 90 W-hr/lb are predicted.

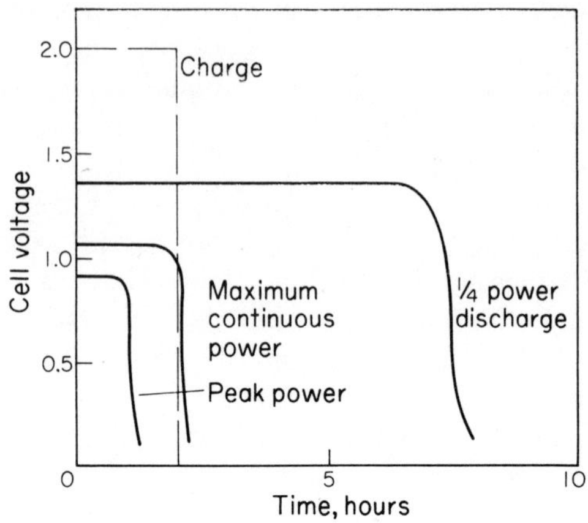

Fig. 22 Cell voltage characteristics.

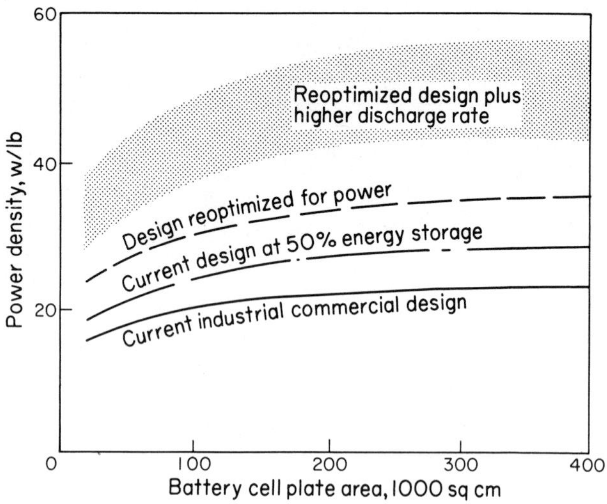

Fig. 22(a) Power density of the General Atomic zinc-air cell may be improved by halving energy storage capacity or by optimising for power.

CADMIUM-AIR CELLS

Cadmium-air cells have a special attraction, because the cadmium electrode is a highly developed component of the nickel-cadmium battery; it is well characterised, with a proven reliability. It is reasonable to assume that the long cycle life of the cadmium electrode can be retained in the cadmium air configuration, provided CO_2 can be kept out of the cell, and provided also that high capacity cadmium electrodes can be developed to take advantage of the infinite capacity of the air electrode.

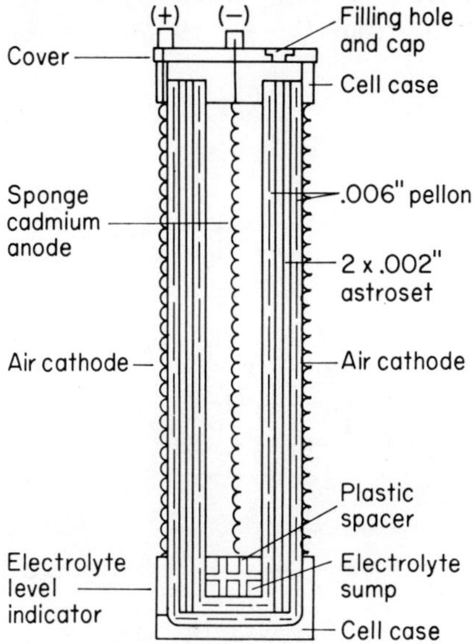

Fig. 23 Unit cadmium-air cell.

The U.S. Army Electronics Command Laboratories[10] have described single cells which deliver 40–50 W-hr/lb at the five-hour rate of discharge. The cells used hydrophobic air electrodes of similar construction to those used in zinc-air cells, with various types of separators. Sponge cadmium anodes were produced by reducing a compact of cadmium oxide, containing nickel oxide and an inert extender (titanium oxide). Each of the anodes was placed between a pair of air electrodes.

Shortcomings of these cells derived from loss of capacity of the anodes, shorting by cadmium penetration of the separator, poisoning of the air cathode by a soluble cadmium compound, loss of catalytic activity by the air electrode, and water loss by evaporation.

Loss of cadmium electrode capacity can be prevented by addition of Fe_2O_3 or TiO_2 extender, and by complete removal of CO_2 from the air stream. Carbonate con-

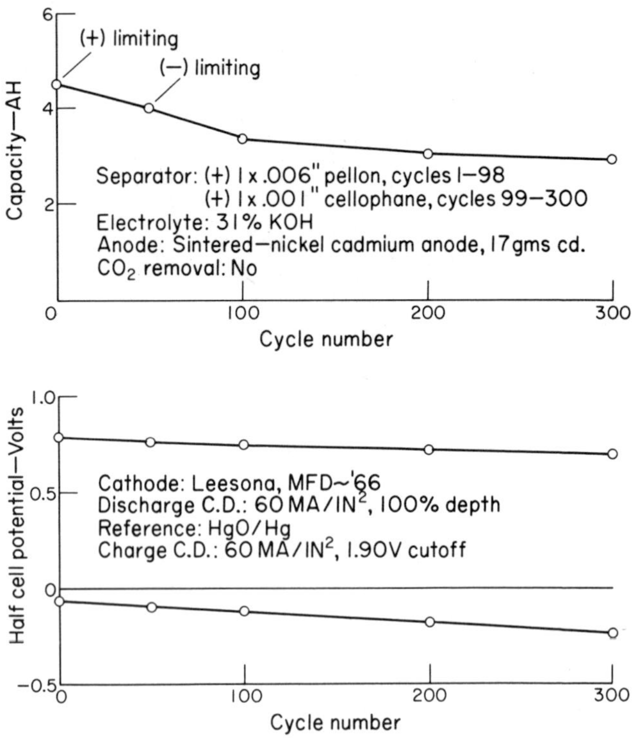

Fig. 24 Performance of cadmium-air cathodes (bifunctional). (CO_2 exposure; 31 per cent KOH.)

tamination also promotes cadmium penetration of the separator; this can be eliminated completely by removing CO_2 and strengthening the separator. Air cathode poisoning was overcome by saturating the electrolyte with zincate or aluminium ions. Deactivation of catalyst and evaporation are problems not yet solved.

Single cells (shown schematically in Fig. 23) incorporating all these modifications have been operated for 300 charge/discharge cycles, at an energy density of 50 W-hr/lb. The cell voltage and capacity history is shown in Fig. 24. However, widespread commercial application is not anticipated because of the high cost and limited supply of cadmium, and because of the severe technical problem of complete elimination of CO_2 from the air supply. The latter is a more critical factor in the cadmium-air system than in other metal-air systems.

SODIUM-AIR CELLS

Atomics International[(11)] have investigated a sodium-air rechargeable system. Sodium is not compatible with an aqueous electrolyte unless it is heavily amalgamated with mercury, when reasonably efficient electrodes can be obtained.

However, if the amalgam is made too rich in sodium, direct reaction between sodium and the water of the electrolyte, producing hydrogen, will occur quite rapidly, causing a drastic reduction of the cell's efficiency. It is found in practice that an amalgam containing 2 per cent sodium represents the maximum concentration which can be tolerated. This means that, although sodium-air has an extremely high theoretical energy density, the weight penalty imposed by the mercury destroys the attractiveness of the system.

Atomics International operated the sodium-amalgam air cell with a relatively small amount of sodium amalgam, and the sodium content was replenished by placing the electrode in contact with another cell containing a molten salt electrolyte, and a liquid sodium anode.

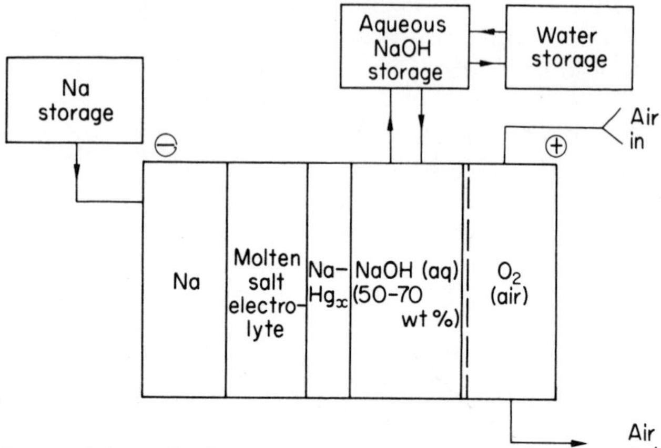

Fig. 25 Schematic diagram of a sodium-air cell.

A schematic diagram of the cell is shown in Fig. 25. The cell consists of a sodium compartment, a molten salt electrolyte matrix, an amalgam electrode, and aqueous NaOH electrolyte compartment, an air electrode and an air compartment. On discharge, sodium is supplied continuously, and is deposited on the molten-salt side of the amalgam electrode. Sodium migrates across the thin stationary amalgam electrode and is made available for the reaction with oxygen in the second cell, where NaOH, the overall reaction product, is formed. Aqueous NaOH is kept in solution by circulation through an external reservoir and is diluted by the addition of water.

On recharge, NaOH solution is electrolysed, depositing sodium on the amalgam electrode, and evolving oxygen on the air electrode. Water is evaporated in order to keep the NaOH concentration constant. Sodium migrates across the amalgam electrode, and is then plated out at the sodium electrode. The two sub-cells operate at the same temperature, in the range 130–160°C; the composition of the molten salt system, which melts at 127°C, has not been disclosed. The air electrode, which contains platinum as a catalyst, is a hydrophobic structure; the amalgam electrode is housed in a thin porous supporting structure.

Cells of up to 2·0 W have been operated for up to 150 hours, at about 50 mA/cm^2 current density. Energy densities of around 200 W-hr/lb have been predicted for complete systems, compared with a theoretical figure of the reactant-energy density of 930 W-hr/lb. This is clearly an attractive concept, but a more experimental research is required for the development of a commercial system.

6. Applications of Metal-Air Batteries

The greatest advantage of the metal-air battery over conventional batteries lies in its great saving in reactant weight. High capacity batteries can be made with a low overall weight. Metal-air cells are thus an attractive proposition for applications which require *portable* power in relatively large amounts.

The electric vehicle is a case in point. Gasoline-powered vehicles use very highly developed engines which are relatively inexpensive and have a high energy-density, but they are noisy, have high maintenance requirements and cause atmospheric pollution. "Conventional" electric vehicles, using lead-acid batteries, are expensive, extremely heavy and therefore underpowered, but are reliable, long lived, clean and silent. In order to compete with the gasoline engine, the electric battery must achieve an energy density of 50–150 W-hr/lb, and be able to deliver power at a rate of 35–95 W/lb. Table 9 shows estimated power density and energy density requirements suggested in a U.S. report (Ref. 12). Many metal-air systems already promise to achieve these figures.

The cost of "metal fuel", however, is too high for consideration to be given to its use in a primary battery for vehicle propulsion. Secondary batteries still present many development problems; also, they require a fairly long charging period which compares very unfavourably with the time necessary to fill a gasoline fuel tank. The use of mechanically rechargeable cells for vehicle propulsion has been suggested, spent anodes being returned to a central service station for processing in bulk, either electrically or chemically using a hydrocarbon fuel source. The mechanical problems of "replating" a multi-cell vehicle battery quickly at a service station appear to be too great at the present time.

*Table 9 Energy Density and Power Density Requirements for Various Electric Vehicles**

	ENERGY DENSITY W-hr/lb	POWER DENSITY W/lb
Family Car	135	94
Commuter Car	41	46
Delivery Van	50	55
Taxi	96	45
Bus	81	36
Metal-Air (Predictions)	50–80	30–40
Lead-Acid (Available)	15–20	20–30
Petrol Engine (Available)	—	65–165

* Source—Reference 12.

The problems inherent in the development of a rechargeable metal-air battery for vehicle propulsion are therefore formidable. Metal-air systems are being considered for transport applications on a long-term basis, but greater attention is being paid to alternative applications which show more immediate promise. An example is to be found in the development of portable power systems for military and commercial communications devices. Mechanically rechargeable zinc-air batteries are already in use for military radio transmitter-receivers and radar sets, which require relatively high power levels coupled with portability. In addition, commercial "walkie-talkie" tranceivers (pocket paging sets) are powered by relatively high power batteries, and light-weight zinc-air cells could find an important outlet in this application.

Electric power tools and domestic appliances are at present operated from the mains electricity supply, necessitating a cumbersome lead. Some battery operated appliances have appeared on the market, using secondary nickel cadmium batteries, but these tend to be heavier and bulkier than their mains-driven counterparts. A metal-air battery would provide a high-energy power source for these devices with a lower weight than nickel cadmium, because the advantages of metal-air systems can be exploited in uses which need relatively high capacity and high rate batteries. Primary metal-air cells, however, will be rather more expensive in operation than an electrically rechargeable nickel cadmium cell, and the preferred choice would be a simple secondary (electrically rechargeable) metal-air battery. Until such a system is developed, we may see a cheap mechanically rechargeable aluminium air cell being developed for this type of application. In general, all portable electrical devices which use electric motors are potential users of a high capacity, light weight metal-air battery. Increasing numbers of toys are being produced with electric motor drive; here again, a metal-air primary or mechanically rechargeable battery system would prove attractive.

Emergency portable equipment often requires high rate batteries or motor generator sets to provide instant power. Such items as emergency lighting in accident or disaster areas out of doors, warning beacons, traffic signals, etc., could make good use of a lightweight portable metal-air battery.

Since more energy can be stored in a primary metal-air battery than in other batteries of a given weight and size, it is likely that "long life" primary metal-air batteries will be used in portable radios, flashlights and other everyday devices.

CONCLUSIONS

Metal-air batteries are not new—zinc-air systems have

been manufactured and used for many years as primary batteries. Recent developments of high rate air electrodes in the fuel cell field has rekindled interest in metal-air batteries, and during the period 1965–1970 a very considerable amount of development work was done. Zinc-air has emerged again as the most promising system. Mechanically rechargeable zinc-air units have been designed and developed for military use in portable electronic equipment, and such units are commercially available.

The recharging procedure for mechanically rechargeable batteries is slightly inconvenient to the "man in the street", and it is likely that primary metal-air batteries will prove more attractive for commercial use for typical everyday electrical devices where battery weight is a significant factor. At the time of writing (1970) no miniature commercial portable primary metal-air battery of the "dry-cell" type is available, though larger static zinc-air batteries have been available for many years.

The development of a lightweight electrically rechargeable battery presents a very real challenge to the scientist and engineer, and much research and development work is being carried out. Many metal-air systems are under investigation. The disadvantages of the air electrode tend to outweigh the intrinsic advantage of having an infinite storage capacity in the atmosphere. At the time of writing, several laboratory models of commercial secondary metal-air batteries have been demonstrated, but none has yet become available commercially.

APPENDIX

	TYPICAL TEMPERATURE, °C	RESISTIVITY OF ELECTROLYTE, ohm-cm	CELL VOLTAGE AT DISCHARGE, V	CURRENT DENSITY, A/cm²	SPECIFIC POWER W/lb	SPECIFIC ENERGY, W-hr/lb	SHELF LIFE, DISCHARGED	OPERATING LIFE, CYCLES
Lead-Acid	−40–50	1·53	2·1–1·46	0·010	3–14	8–14	None	10–400
Nickel-Iron	0–40	1·96	1·3–0·75	—	3–18	14–16	Years	100–3000
Nickel-Cadmium	−40–60	1·96	1·3–0·75	0·010	3–20	16–18	Years	100–2000
Silver-Zinc	0–40	1·96	1·55–1·1	0·43	11–68	36–45	Years	100–3000
Silver-Cadmium	−40–60	1·96	1·3–0·8	—	9–30	23–27	Years	500–1100
GM Lithium-Chlorine	650	0·17	3·40	1–3	36–68	150	—	—
Electrochimica Lithium-Metal Halide	25	—	2·1	—	—	19	—	—
Ford Sodium-Sulphur	300	5 (Solid)	2·08–1·76	0·7	100	136	—	—
AI Sodium-Air	130	3·58	2·3	0·07	41	159	—	—
G A Zinc-Air	25	—	1·4	—	27	91	—	—
ANL Sodium-Bismuth	540–580	0·44	0·8–0·4	0·5–1	36	18	Years	>500
ANL Lithium-Tellurium	450–480	0·26	1·79–1·67	2–5	127	82	Years	—

GA = General Atomic   AI = Atomics International

GM = General Motors   ANL = Argonne National Laboratory

* Taken from Reference 5.

OUTLOOK FOR VARIOUS BATTERIES

SYSTEM	ENERGY DENSITY	POWER DENSITY	RECHARGEABILITY OR REFUELABILITY	SERVICE LIFE	MATERIAL AVAILABILITY	COST	OTHER PROBLEMS
Lead-Acid	Poor	Good	Good	Good	Excellent	Excellent	—
Nickel-Iron	Poor	Fair	Good	Excellent	Good	Fair	—
Nickel-Cadmium	Poor	Excellent	Excellent	Excellent	Poor	Poor	—
Silver-Zinc	Fair	Excellent	Poor	Poor	Poor	Poor	—
Zinc-Air	Good	Fair	Good	Fair (?)	Excellent	Good in Production	Complexity maintenance (?)
Magnesium-Air	Excellent	Fair	Fair (?)	?	Excellent	Excellent in Production	Refueling is mechanical
Lithium-Organic Electrolyte	Excellent	Poor	Fair to Good	Fair (?)	Good	Good in Production	—
Sodium-Sulphur	Good to Excellent	Excellent	Excellent	Excellent	Excellent	Excellent in Production	Safety, high temperature, startup
Lithium-Chlorine	Excellent	Excellent	Excellent	Excellent	Good	Good	Safety, toxicity, complexity, startup (?)
Lithium-Tellurium	Excellent	Excellent	Excellent	Excellent	Poor (?)	Poor	Temperature toxicity, startup (?)

PRESENT BATTERY STATUS

Present Battery Status

Prototype Batteries

TYPE	COMPOSITION (CHARGED) +CATHODE	−ANODE	ELECTROLYTE	AVERAGE ENERGY w-hr/lb THEORETICAL	ACTUAL	w-hr/cu. in.	MAXIMUM POWER DENSITY, w/lb	CELL POTENTIAL, V OPEN	DISCHARGING	CELL LIFE	REMARKS	MANUFACTURER OR DEVELOPER
Zinc-Nickel-S*	Ni	Zn	KOH	220	40–50							Yardley Electric
Zinc-Air	Air	Zn	KOH	464	50–60 to 100	2–3	30 (40 est.)	1·65	0·9 1·2 1·4	Several hundred cycles	Several approaches to avoid dendritic zinc	GE, General Atomics (Gulf), Yardney, ESB
Zinc-Air-P**	Air	Zn	KOH	464	150		20 (40 est.)	1·65	0·9–1·2 1·3			Leesona-Moos
Magnesium-Air-P	Air	Mg	Aqueous solutions of NaCl, CaCl, LiCl, M_5Cl		500 with local H_2O refills							GE
Sodium-Sulfur-S	S	Na	Ceramic	345	84–100 est. to 150	4 est. 8	100 est. to 200	2	1·75	Indefinite ?	Operates at high temperature (250–300°C), self-maintaining once brought to operating temperature	Ford
Sodium-Air-S	Air	Na	NaOH	930	160–215 (4 hr, discharge	5	40–55 (4 hr discharge	2·60	2·3–2·4	Indefinite ?	2-step process involving complex sodium amalgam—operates at 130°C	Atomic International (Northern American Aviation)
Sodium-Air-P	Air	Na	NaOH	930			~40	1·9–2·0	1·4		Sodium-mercury amalsam, ambient temperature	Western Reserve
Calcium Air-P	Air	Ca	KOH									Yardney
Lithium-Moist Air-S	Air + H_2O	Li	Nonaqueous	2566								Globe-Union
Aluminium-Air-P	Air	Al	KOH		240–400		30 av. 48–75 peak	2·7	1·1		Power controlled by adjustment of liquid level	Zaromb Research Corp.
Iron-Air-S	Air	Fe		650							Proprietary development	Westinghouse

Dry tape-P	Halosem systems	Li	Organic		200 (ex-hardware)			3·0–3·1			Monsanto
		Mg	Aqueous		100 (ex-hardware)			2·0			
Dry tape-P	Ag:O_2	Zn	KOH		25–30						Monsanto
Lithium-Chlorine-S	Cl_2	Li	Lithium-chloride	1200	100 est. to 300	75 est. to 150	3·5	3·9	Indefinite ?	Operates at 650°C	GM
Lithium-Tellurium-S	Li:Te	Li	LiCl-LiF eutectic	270	90	140	1·94	1·67–1·79	Indefinite ?	Operates at 471°C	Argonne National Laboratory
Lithium-Nickel Chloride-S	$NiCl_2$	Li	Propylene carbonate or K-phosphofluoride	435	est. 100	Low <20	2·50			Room temperature	Gulton
Lithium-Nickel Fluoride-S	NiF_2	Li	Same as above	620	90–100	Low <70	2·9			Room temperature	Gulton
Lithium-Silver Difluoride S	AgF_2	Li	Butyrolactone, KPF_4	678		Low			Has undergone 50 cycles at 90% depth of discharge	Room temperature	Whittaker
Lithium-Silver Chloride-S	AgCl	Li	Li salts in PC, other organics	231	30 est. 90	Low		2·3–2·8	50–200 cycles	Room temperature	Lockheed, Electrochimica
Lithium-Copper Fluoride-P	Cuf_2	Li	Li salts in PC, other organics	746	57–80 110 est. 200	Low	3·6	2·3–2·8		Room temperature	Lockheed, ESB Electrochimics
Lithium-Copper Chloride-S	$CuCl_2$	Li	Li salts in PC, other organics	503	25 est. 175	Low	3·1	2·3–2·8		Room temperature	Mallory Electrochimica

*Secondary
**Primary
Existing Batteries

TYPE	COMPOSITION (CHARGED) +	−	ELECTROLYTE	AVERAGE ENERGY DENSITY w-hr/lb THEORETICAL	ACTUAL	w-hr cu in	MAXIMUM POWER DENSITY, w/lb	CELL POTENTIAL OPEN	DISCHARGING	CELL RATED CAPACITY, amp-hr	DISCHARGE RATE amp	CAPACITY % RATED 80°F	0°F	LIFE IN OPERATION CYCLES	FLOAT	AVERAGE COST, RATED kw-hr CURRENT SLI SIZE: 12V 50 amp hr	MANUFACTURERS (SEE FOOTNOTES)
Lead-Acid	PbO_2	Pb	H_2SO_4	112	8·13	1·13	32	2·14	2·1–1·46	60	10 25 50 100	100 87 63 39	54 31 18 —	Automotive 10–400, industrial 1600	to 14 years	30–90	a, b, c, d, e, q, r
Nickel-Iron	NiO_2	Fe	KOH	~115	12·13	0·98–1·05		1·34	1·3–0·75	75	15	100	67	Up to 2000	8·20 years	100–150	a, p
Nickel-Cadmium	NiO_2	Cd	KOH	107	15·90	0·7–1·3	40 bipolar 150 to 1000	1·34	1·3–1·00 (+) in series	75 (up to 850)	10 25 50 100 1000	100 97 94 82 20	86 82 72	100–200 300–3,000	8–14 years	1700·00 250·00–350·00	b, c, f, g, h, l, l, p, s, t
Silver-Zinc	Ago	Zn	KOH	254	50–60 (low rate) maximum ~90	1·8–3·0	~150	1·86	1·55–1·1	60	10 25 50 100	100 97 92 84	92 79 75 69	100–300	1–2 years	650·00–900·00	a, b, c, j, k, l, m, n, u
Silver-Cadmium	AgO	Cd	KOH	134	20·25 low rate maximum ~40	0·4–2·6	65	1·34	1·3–0·8	60	6 60	100 95	96 80	300–1500	2–3 years	900–1000	a, k, l

List of Battery Manufacturers

a. Electric Storage Battery Co.
b. Gould National
c. Globe Union
d. Autolite (Ford)
e. Delco-Remy (General Motors)
f. Sonotone
g. Gulton Industries
h. Union Carbide
j. General Electric
k. Yardley Electric
l. Eagle-Picher
m. Aerojet-General
n. Telecomputing Corp.
p. McGraw-Edison Co.
q. General Battery and Ceramic
r. Eltra Corp.
s. Eastman Kodak
t. Saft Corp. of America
u. Frank Corp. Co.

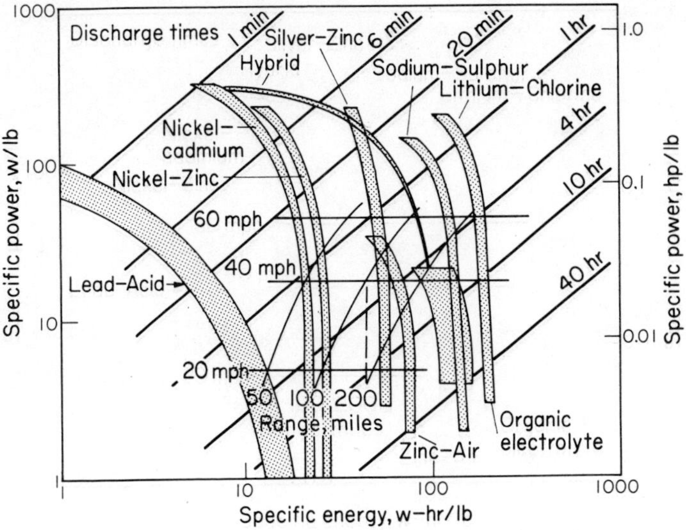

i-c and e-c engines have characteristics in the top right hand corner. Hybrid is a nickel-cadmium/lithium-organic pairing.

BIBLIOGRAPHY

SUGGESTIONS FOR FURTHER READING

There is no textbook dealing with metal-air batteries as its main subject. The reader may be interested in acquiring background to batteries generally, when he will find the following books useful:—

1. G. W. Vinal, *Primary Batteries*. Wiley, 1950.
2. G. W. Vinal, *Storage Batteries*. Wiley, 1951.
3. S. U. Falk and A. J. Salkind, *Alkaline Storage Batteries*. Wiley, 1969.

REVIEWS ON METAL-AIR BATTERIES

4. H. G. Oswin in *Performance Forecast of Selected Static Energy Conversion Systems*, p. 397. AGARD Propulsion and Energetics Panel, 1967.
5. *S.A.E. Journal*, **76,** No. 12, p. 61 (1968).

ZINC-AIR—MECHANICALLY RECHARGEABLE

6. S. M. Chodosh, B. Jagid and E. Katsoulis, in *Power Sources 2*, p. 423. Ed. Collins, Pergamon, 1968.

ZINC-AIR—SECONDARY

7. D. V. Ragone, in *Power Systems for Electric Vehicles*, p. 225, Report PB 177, 706, National Center for Air Pollution Control, 1967.

MAGNESIUM AIR

8. W. N. Carson Jr. and C. E. Kent, in *Power Sources 1966*, p. 119. Ed. Collins, Pergamon, 1966

ALUMINIUM AIR

9. S. Zaromb, in PB 177, 706, p. 255.

CADMIUM AIR

10. O. C. Wagner, *J. Electrochem. Soc.*, **116,** No. 5, p. 693 (1969).

SODIUM AIR

11. L. A. Heredy, H. L. Recht and D. E. McKenzie in PB 177706, p. 245.

ELECTRIC VEHICLES

12. J. H. B. George, L. J. Stratton and R. G. Acton, *Prospects for Electric Vehicles.* A study of Low pollution Potential Vehicles—Electric. Arthur D. Little Inc. Report C. 69260, May, 1968.

INDEX

Air electrode 8, 17–20, 49
Air electrode mechanism 17–18
Air electrode structure 19–20
Air Supply 33, 34
Aluminium 26
Aluminium-air cell 42–45
Atomics International 63–65

Cadmium-air cell 61–63
Catalyst 7, 19
Cell capacity 54
Communications devices 67
Corrosion 43, 45, 46, 50–52, 57

Dendrites 56
Dry cells 14, 69
Dry tape cell 30

Electric vehicles 66–67, 75
Electrolyte 26–27
Energy Conversion Ltd. 38, 40, 41, 53
Energy density 42

Fuel cells 16, 30

General Atomics Co. 58–60
General Electric Co. 45, 46, 47
General Motors Corporation 57

Intercell separator 33–35

Joseph Lucas Ltd. 58–60

Leclanche cell 10, 49, 52–53
Le Carbone Co. 54, 55
Leesona Moos Laboratories 32, 35, 36

Magnesium 26
Magnesium-air cell 45–48
Manganese-alkali cell 49
Mechanically rechargeable cell 15, 28, 32–48
Mercury 50, 63–64
Metal electrode 21–25

National Carbon Co. 12, 53

Oxygen electrode 7, 8

Power tools 67
Primary cells 28, 49–55

Recharging techniques 37–40, 57

Secondary cells 28, 56–65
Silver-zinc cell 56
Sodium-air cell 63–65

Third electrode cell 57, 58

Union Carbide Co. 10

Water loss 35, 49, 52
Waterproofing 19, 20

Zaromb Research Corporation 45
Zinc electrode 21, 25, 32
Zinc-air cells 32, 38, 49–55, 56–60, 69
Zinc-oxygen cell 32